Weather and Climate

The Advancement of Science Series

Editor: Richard Carrington

Weather
and Climate

R. C. Sutcliffe, F.R.S.

W · W · Norton & Company · Inc ·

New York

Library of Congress Catalog Card No. 66-11653

Printed in the United States of America

1 2 3 4 5 6 7 8 9 0

Contents

Illustrations

Illustrations (contd.)

Acknowledgements

The author wishes to thank the following for supplying photographs for use in this book: Plates 1–11, R. K. Pilsbury; Plates 12–14, H.M.S.O.; Plate 15, U.S.A. Weather Bureau

Preface

THIS is not a textbook on meteorology, neither a general introduction nor a formal course, but it has a serious purpose and that is to explain to the general reader what it is that meteorologists are doing and trying to do. Much is heard of the two cultures and of science and technology becoming intolerably specialized and sophisticated with concepts often unintelligible to the non-scientist. And yet of necessity they exist, as indeed do literature, art and music, in an economic and political environment where the most far-reaching decisions affecting them rightly rest in the hands of non-specialists who unfortunately may not adequately understand the issues before them. For my own part experience in presenting scientific considerations to non-scientists, civil and military, has rarely left an impression of great difficulty in communication and it could be that the gravity of the present situation has, in this respect, been much exaggerated, but it is still the duty of all specialist groups, which in a democracy draw their resources ultimately from the public purse, to avoid the arrogance of specialized knowledge and to try to present a fair account of their achievements, objectives and requirements in terms intelligible to a large proportion of educated citizens. This I have attempted for weather science, assuming on the part of the reader no more competence than seems to be possessed by most of the intelligent people one meets. It is too much to hope that I have not betrayed an exaggerated sense of the interest and importance of the subject, for the unbiased specialist is a poor creature almost by definition, but I have tried to give an accurate and sober account of some aspects of a science and profession which holds a modest but old-established position in the world. If the account is not both interesting and intelligible to a wide public the fault, we may be sure, lies neither in the subject nor in the reader. That such accounts can be and should be written I have no doubt.

R. C. SUTCLIFFE

Chapter 1

Introduction

WHEN we begin to write about weather and climate we embark upon the story of our natural home, which has been our dwelling in one continuous stream of life these thousand million years, and for the last million or so has been explored and exploited by the conscious mind of man. It is then not unreasonable to suppose, indeed it could hardly be otherwise, that the problems presented by weather, by wind and rain and warmth, were amongst the earliest to force themselves on consciousness and that in a historic sense meteorology lay at the foundation of physical science. It was, and is, a difficult science to reduce to its basic principles and so to present as a deductive structure, and it was another of the environmental sciences, astronomy, the very limited positional astronomy of the solar system, which was first illuminated by the light of Newton's genius. But throughout the history of modern science dating from that time it was significant that 'natural philosophy' was almost a synonym for physical science. The dual interest of the scientist in the natural world of phenomena and in the basic principles which explain them, which identify the natural with the rational, was accepted generally and did not begin to lose its validity until – with the tremendous success of experimental laboratory physics of the late nineteenth century – the applications of basic physical theory were largely diverted from the natural macro-environment of man to the essentially simpler physical systems which he invented for himself and learned to construct and control. It is of course easy to overstate the case but it can hardly be denied that physics of the present century has become preponderantly the basic science of engineering, of the completely revolutionary electrical, electromagnetic, electronic, atomic, and nuclear engineering, and of chemical engineering with its ever-expanding range of new materials created by man. What we call progress, and I believe it is rightly so called, has come not so much by studying and understanding our complex natural environment and adapting ourselves to its behaviour, but by creating an artificial technological and controlled environment. It could be argued that the difference is of degree, not of kind, and that invention and artefact have always been the essential material factors in civilizations, but the modern application of true science is unique. The classical scientists were philosophers not engineers and even our own industrial revolution owed little to science, whereas in recent years the scientists have become the back-room boys, the boffins, the brains behind the business, and 'natural science', as the term is most naturally understood, has suffered astonishing neglect.

That the neglect is extreme is illustrated by the very curious fact that a student entering the science stream will go through school and university studying physics and mathematics and have no more idea than an arts student of the significance of an earthquake, an ocean current, or a monsoon wind, to take illustrations from three of the environmental sciences. Actually we must go to schools of geology and geography to find students who have been asked to direct their minds to the natural phenomena on earth, and these schools have had little chance of all-round achievement when talent in physics and mathematics has been selectively skimmed as it rose to the top and diverted ultimately to the production of radio equipment and nuclear power-stations. To decry schools of geography has been a commonplace reaction of the physicist, presumably because they did inadequately many things which the physicist had on his conscience and should have been doing himself.

There is a ready explanation in that universities – protest as they will – must adapt themselves to meet vocational demands and that seismologists, oceanographers and meteorologists have been needed comparatively little in an industrial world. I have no doubt that this is the true explanation although it puts in rather a poor light the claim of the universities to pursue knowledge for its own sake and in some ways it would be more charitable to excuse the neglect on the grounds that these 'natural' sciences are exceptionally difficult and call for more resources than a university department can easily command. We must, however, accept that we live in a competitive world, that university departments cannot easily expand without students and that students look to the future. Fortunately there has always been a number of university teachers who have been attracted by the earth sciences and have kept the spirit alive, although in rather specialized fields. Meanwhile national institutes for geophysical subjects, meteorology, oceanography, hydrology, agricultural science, and the like, have maintained a healthy activity which is in a position to expand to meet the demands now increasing for advancement in these fields.

To some extent perhaps the current surge of interest is not quite spontaneous and may yet lose its force, for it has to be stimulated as much by engineering invention as by the spirit of scientific research and the pursuit of knowledge. The rocket, the earth satellite, and the space vehicle are splendid constructions which – apart from their military applications and sheer entertainment value, the two great considerations of the age – provide a means for obtaining information of much scientific interest at tremendous cost and scientists have not been slow to take advantage of the facilities which the public purse provides with less pure intent. Whatever may be the ultimate value of space research and even in the most unlikely event that it should founder on economic rocks, we may have faith that discoveries about our atmosphere and oceans will be too close to our livelihoods not to pay a dividend in terms which governments will understand. 'There is a tide in the affairs of men which, taken at the flood, leads on to fortune', is a thought which the meteorologists and oceanographers should readily understand.

Meteorology is not a fundamental physical science, that is to say it is not concerned to develop the basic laws of nature or the fundamental truths of space, time, and matter, but it is evidently a pure science properly to be studied in its own right as a branch of knowledge and will in this book be presented in that light. Yet, it is eminently a useful science with applications of economic value in the widest range of human interests, industry and agriculture, shipping, aviation, and all kinds of communication, through to sport and recreation. Probably for most people meteorology is identified with weather-forecasting, and specifically with the weather-forecasts which are provided as a public service by a benevolent government through press and broadcasting. This application is of primary importance, but it is not the only one and if we were to omit the service provided for aviation, a specialized application, we should be in doubt whether it were in economic terms even the most valuable. Ordinary weather-forecasting is concerned only with the current weather and the expectation for a short period ahead – a few days at most – in terms of the elements, wind, temperature, clouds, visibility, and precipitation, little else. If meteorologists had no skill in distinguishing one day's weather from the next and could therefore provide no advice on that score their services would still be almost as necessary as ever as advisers on how the atmosphere behaves in general, how it has behaved in the past and how meteorological conditions might therefore be expected to affect any future activity of interest. This might be called climatology provided that we include not only the elements of conventional climate but also every other kind of factor or phenomenon anywhere in the atmosphere and anywhere in the world. An organization such as a national Meteorological Office or Weather Bureau will deal in the course of a year with thousands, even tens of thousands of inquiries not concerned with forecasting. To list a few of these is to be in danger of failing to indicate the diversity; but one may, by way of illustration, consider air movement alone. There is wind as ordinarily understood, the knowledge of which is essential for wise planning of all kinds of structures. Before any novel enterprise is undertaken, it may be a skyscraper, an exposed bridge, a motorway over hilly terrain, the effects of normal and possible extremes of wind must be assessed often on the basis of inadequate observational data. Appeal will be made to theory as well as to climatological data, perhaps to model tests in wind tunnels as well as to comparison with other structures; the inquiry, with the meteorologist as an essential collaborator, may be long and exhaustive. But ordinary wind is not the only kind of air motion and inquiries may refer to diffusion of smoke from factory chimneys and its concentration in different places and different circumstances. The meteorologist may then play the rôle of design consultant or it may be of expert witness in a court of arbitration basing his opinion and advice on a wealth of empirical and theoretical knowledge on atmospheric turbulence, a subject which enters in another guise, and in a critical way, to the design of aircraft. Examples could be multiplied and when, as well as air movement, we think of the many ways in which an enterprise may be affected by other factors: by temperature – it may be heat or cold, by humidity, by atmospheric pollutants, by radioactivity, by sunshine, perhaps by ultraviolet radiation to pick out a special point, by visibility, by rainfall – perhaps heavy raindrops on a supersonic aircraft, by ice or snow, hail or thunderstorm, it will be evident that the meteorologist, the atmospheric physicist if the term is preferred, has innumerable points of contact with the affairs of the world without necessarily being prepared to hazard an opinion on tomorrow's weather.

In the course of this book there will be little room to discuss applications of the science and to do so is not the intention, but, in making the general case by way of introduction, I have in mind the further justification of writing a serious specialist work for the non-scientist. It is desirable that a proportion of those who pay for a public scientific service should be aware of the problems that are set, how they are being attacked, and how they may be usefully applied. It is the established tradition in the profession of meteorology to take the public into one's confidence. We have no trade secrets, no deliberate professional obscurantism such as the ancient professions have built up around themselves. Perhaps no scientist is more regularly and more unmercifully laboured with questions of his trade wherever he may be than is the meteorologist; it is an occupational hazard which, fortunately, he takes in good part. He usually enjoys 'talking shop', which could be the title of this book.

The past neglect and the growing awareness of meteorology in schools of science, combine in further encouraging the writing of this book for it may be of interest not only to the general reader, the layman to whom it is mainly addressed, but it may – if it is fortunate enough to reach them – be congenial and easy introductory reading to that large body of people trained in physics who have never given their thoughts to the subject although they are so well qualified to do so.

Chapter 2

Troposphere, Stratosphere and Beyond

THIS book is about weather and climate as ordinarily understood and it will rarely be necessary to consider the behaviour of the atmosphere beyond the highest level reached by the clouds, say 10–15 kilometres, by commercial aircraft, say 20 kilometres, or at most the limit of balloon ascents, say 40 kilometres. The atmosphere continues, however, in ever more tenuous form outwards for some hundreds of kilometres and if its behaviour has little effect upon our weather it is of remarkable diversity and scientific interest. Being held to the surface of the earth only by the force of gravity, the air at any level is under pressure by the weight of the air above and, consequently, the pressure and the air density decrease steadily with height. The rate of decrease with height is not uniform being dependent upon the temperature and the composition of the air but as a rough rule the density may be divided by a factor of 10 for each 20 kilometres of height. On this basis the density at 200 kilometres is less than that at the ground by the factor 10^{10}, ten thousand millions, which for many purposes could be regarded as defining a complete vacuum. The number of molecules of gas to the cubic centimetre is, however, still enormous, some thousands of millions, and on the scale of geophysics there is still an atmosphere with observable and important properties.

The steady decrease of density with height, more or less as described, is the inevitable result of hydrostatic compression by the force of gravity but the variation of temperature with height, far from being steady, is altogether remarkable. When it became firmly established from observations on mountains and in manned and free balloons that the air became steadily colder as the altitude increased, scientists were very ready to generalize and to assume that the cooling went on indefinitely to the limit of the atmosphere.

This was the general belief until in 1899 the Frenchman Teisserenc de Bort, announced to an astonished and even incredulous world that his sounding balloons had reached heights above which the temperature decreased no further. Sir Napier Shaw, the leading British meteorologist of the early decades of the century, called this the most surprising discovery in the whole history of meteorology. It is now known that the limit comes at a varying height and temperature, averaging 10 kilometres and −45°C in middle latitudes but much higher and colder, 17 kilometres and −80°C, within the tropics. The terminology to distinguish the two regions, the troposphere below and the stratosphere above, separated by the 'tropopause', was then introduced and meteorologists settled down comfortably once more to their two-storey structure which survived for another generation. Instead of cooling continuously with height, the atmosphere within the stratosphere, it was now credibly inferred, retained much the same temperature at all heights, although colder at lower than at higher latitudes: the stratosphere was sometimes called 'the isothermal layer'.

Then in the nineteen-twenties the two-storey model had in turn to be abandoned. It became known that there must be at least one more storey, a warm region, almost as warm as near the ground, in the neighbourhood of 50 kilometres. The evidence came first from observations of the heights of meteors (shooting stars) and later from much more conclusive calculations on the audibility of explosions at great distances. This nice piece of work is especially associated with the names of Lindemann and Dobson of Oxford, the same Lindemann who, later as Lord Cherwell, was to become famous during World War II as personal scientific adviser to Winston Churchill. Professor G. M. B. Dobson remained faithful to his corner of science and gained world-wide fame for his studies of atmospheric ozone – another romantic episode in meteorological exploration, of which more will be said later.

The high temperature at 50 kilometres was a scientific inference made long before instruments could be sent to such heights. There was no other reasonable explanation of the way in which sound waves from distant explosions could be bent down towards the earth to be heard at distances of 200 kilometres or more after skipping inaudibly over a nearer zone, as illustrated in Fig. 1. This technique was not, however, capable of giving information from above the warm layer at about 50 kilometres and the discovery of what we might call the fourth storey came mainly from

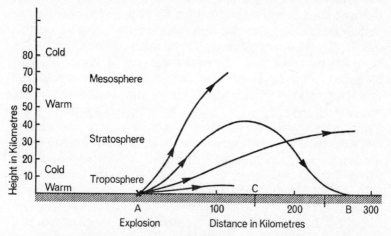

Fig. 1. Sound waves from an explosion at A are refracted downwards, and the explosion may be heard at the distant point B although inaudible at the immediate place C. This phenomenon led to the discovery of the warm layer of the atmosphere at heights of about 50 km.

observations by rockets: there was another cold layer with a minimum of temperature, at a height of about 80 kilometres, even colder than the tropopause, and this was surmounted by warmer air once more. It now seems to be firmly established that there is no further setback and that the temperature continues to rise beyond 100 kilometres to remarkably high values estimated as above 1000°C. These temperatures also have been inferred mostly by very indirect evidence, in particular from careful observations of the orbits of artificial earth satellites. The variation of the orbit with time depends upon the frictional drag of the atmosphere; the drag depends upon the air density; the density upon the temperature and so, provided the composition of the atmosphere is known, the chain of calculation may be completed.

The up-to-date picture of the average temperature in the atmosphere from the earth's surface until the air merges into inter-planetary space (shown schematically in Fig. 2) has, it now appears, two cold layers sandwiched between three warm ones, an arrangement which was never even guessed at before it was discovered by observation, and which is not entirely easy to explain even when it is known. One curious twist is in the present firm belief that the outer fringe of the atmosphere is intensely hot, whereas early in the century it was thought to be intensely cold; although one should perhaps add that the temperature of the outer billionth part of the air is not a matter of deep concern to those living on the earth – about as important as the temperature of the top surface of the ocean, say a thousandth of a millimetre thick, is to a deep-sea fish. The reference to the sea puts one in mind of the tides and the waves, and it is of interest to mention that outwards beyond some 200 kilometres the earth's tenuous atmosphere is subject to very large rhythmical variations of temperature and density, tides and waves of a kind.

The study of the phenomena and processes in the mesosphere (see Fig. 2) and beyond has in recent years developed into a science in its own right, as distinct from the meteorology of the low atmosphere as this is from the study of the oceans. After all, the density ratio of surface air and water is a mere 1:1000, whereas that between air at 200 kilometres and that at the ground is 1:10,000,000,000. Furthermore, the fluid above 200 kilometres is no longer air as we know it at the ground but is thought to become almost pure oxygen in the atomic form, the molecules having been dissociated under the influence of ultraviolet light from the sun and separated from the relatively heavy nitrogen by gravity. At still higher levels the gas may be almost pure helium and hydrogen.

From about 70 kilometres upwards to about 500 kilometres we have what has been called for a considerable number of years the ionosphere, with layers which for a time were associated with the famous pioneer names Heaviside and Appleton, but have since been granted the doubtful blessings of anonymity with the letters D, E, and F. Here the atmosphere is ionized to such an extent that it becomes effectively a conductor of electricity and an efficient reflector of radio waves without which long-distance

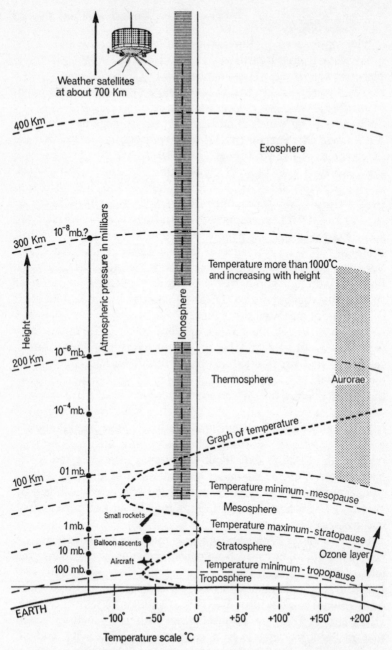

Fig. 2. Structure of the atmosphere up to 400 km height. All weather is concentrated in the lowest 10–15 km – the troposphere.

radio communication in its present form could never have been developed. The study of the ionosphere all over the world, and especially of the disturbances caused by solar flares and associated with 'magnetic storms', has therefore become an important branch of radiophysics.

Ozone For more detailed accounts of the many fascinating phenomena of the outer atmosphere the reader is referred in the bibliography to a number of recent books but no modern account of weather is complete without some attention to the so-called 'ozone layer', quite one of the most curious features. I use the term 'so-called' because the ozone in the air, even where it is most concentrated, accounts for only a few parts per million so that in the ozone layer the air is still much the same mixture of nitrogen, oxygen and carbon dioxide that we have in the troposphere, but the minute quantity of ozone is all important.

The occurrence of ozone gas depends upon photochemical reactions of some complexity. There are three forms of oxygen: the ordinary oxygen we breathe has molecules each of two atoms, O_2; ozone has molecules of three atoms, O_3; and the single atomic form may also exist, O. Now O_2 will absorb ultraviolet light of sufficiently short wavelength and split into two separate atoms, $O + O$. Then O_2 and O being present together very readily combine to form O_3 while O_3 itself is also dissociated by ultraviolet radiation to reform O and O_2. There are other possible reactions, for two oxygen atoms may meet and combine once more to O_2, or O and O_3 may meet to form two molecules of O_2 – and all these reactions go on simultaneously depending on the abundance of the participants and upon the availability of the ultraviolet radiation. The basic theory of chemical and photochemical reactions of this kind is reasonably well known and it is therefore possible to calculate with some confidence how much O_3 and O should be present in photochemical equilibrium in different circumstances and at different heights; but the calculations do not give answers which accord with the observations from the atmosphere. They account very well for the fact that ozone does occur at the upper level where it is found and in much the right concentration but they do not account for the ozone which is found also in parts of the atmosphere where it could not be produced photochemically.

At first sight the result is paradoxical but the explanation came rather readily. There was little reason to doubt the photochemical theory and therefore one looked for the effects of the motion of the air in conveying ozone-rich air into other regions.

By general subsidence of the air or by turbulent mixing, ozone may be brought downwards into the region where very little ultraviolet light can penetrate and there is then nothing to disrupt the gas. Ozone is such a good absorber that only a very small amount is needed to cut off the solar spectrum at the short-wave end (beyond about 3000 Å) with the result that if some of the ozone formed at heights above 40 kilometres can sink below 30 kilometres it will be completely protected and, as calculations confirm, may survive for several months. Then again, if ozone formed in the sunlit regions of the atmosphere can be carried by the winds into latitudes where there is little sunshine, in the extreme case into the darkness of winter near the poles, it may accumulate for months without being destroyed. Destruction does, however, come about quite quickly by chemical reaction whenever ozone comes into the troposphere and encounters various pollutants. To sum up the position, we may say that ozone storage occurs predominantly in the stratosphere below 30 kilometres especially during the winter and at higher latitudes, whereas ozone production is mainly above 40 kilometres during the summer and at lower latitudes. It is the efficiency of the storage out of reach of ultraviolet radiation which mainly determines the distribution at any one time.

The immediate practical importance of ozone lies in its effectiveness in filtering ultraviolet from the sunshine. Without this protective screen life on earth would presumably have evolved differently to provide self-protection from the rays which are damaging to many tissues. The absorbed ultraviolet represents a significant part of the energy of the sunshine, perhaps 6 per cent, and accounts for the high temperatures in the ozone layer. To the meteorologist the importance of ozone also lies in its value as a tracer of air motions and our ideas about the circulation of the air in the stratosphere from one part of the world to another owe a good deal to the need to explain how the ozone finds its way into the polar regions in the winter and spring and how the amount of ozone overhead varies with the synoptic weather conditions.

22

The way in which motions near the ground and in the higher stratosphere are linked together is exemplified by there being much more ozone over depressions than over anticyclones.

The measurement of the total amount of ozone overhead at any time is a skilled observation using a spectrophotometer which measures, in effect, the amount of absorption in a line of the solar spectrum chosen as one which ozone absorbs well but not too well. Instruments, following Dobson's design and costing about £2000, have now been installed in many countries including the Arctic and the Antarctic. There is indeed a world network of ozone-measuring stations – about sixty in all – organized largely by the International Ozone Commission, the observations being collected centrally and published by the Canadian authorities as a voluntary contribution to world science. The newest departure is the development of 'ozone sondes', balloon-borne instruments of various designs which telemeter to the ground data on the ozone concentration as it varies with the height of the balloon. A world network of ozone-sounding stations is clearly on the way.

Chapter 3

Exploring the Free Atmosphere

No small part of the attractiveness of the earth sciences, including geology and geography as well as what we call geophysics, lies in meeting the challenge of exploration, by expeditions of discovery to all parts of our planet and by devising methods of obtaining information from the inaccessible interior of the solid earth, the depths of the oceans or the heights of the atmosphere. We are now said to be moving into the space age, which means pushing the exploration far out beyond our own planet; but the story of the probing of the earth's atmosphere and of the scientific instruments developed for the purpose is mostly quite recent history, not too mundane to be worth the telling.

We should perhaps begin with early days of descriptive observations when explorers, unaided by instruments, brought back accounts of winds and weather, of clouds and fog, of warmth and cold, and meteorology or climatology was established as one of the earliest of sciences. It was not until the seventeenth century that we began to move forward, and then only slowly, into the scientific age of measurement when, with the use of the barometer and the thermometer, meteorology began to assume the character of an 'exact science'. The nineteenth century saw the development of classical physics, of thermodynamics and the concept of energy, and incidentally of the full range of standard meteorological instruments, thermometers, hygrometers, anemometers, rain-gauges and the like which, for the most part, have undergone little modification in the present century. I shall not, however, spend time in going over all this old ground but shall be content with a brief outline of some current techniques for exploring the free atmosphere, which have been introduced wholly in recent decades, some only in the last few years. The items selected for this purpose are the balloon-borne radio-sonde and radar for wind measurements, the weather or storm radar, the meteorological rocket and, lastly, on the common ground of meteorological exploration and space technology, the earth-orbiting weather satellite.

First a few words about the balloon, still the most important vehicle. A simple rubber balloon, nothing more sophisticated than a child's toy balloon, will when inflated with hydrogen rise by buoyancy to 10,000, 20,000 or 30,000 feet before bursting. The meteorologist employs in large numbers quite small balloons, one stage better than the toy costing a few pence, to carry a load of a pound or two. There is no great difficulty in reaching 60,000 feet with regularity with a balloon costing about £2, but higher levels demand envelopes of great uniformity and strength. It will be recalled that the air pressure at the ground is about 1000 mb and as a useful rough rule decreases to 100 mb, 10 mb, and 1 mb at heights of 50,000, 100,000, and 150,000 feet. The hydrogen would naturally expand at these levels to 10, 100, and 1000 times its initial volume and it will readily be appreciated that practical ballooning soon runs into prohibitive difficulties. To attain 100,000 feet with reliability requires at present a balloon costing some £8 which for routine use every day at many upper-air stations begins to add up to a major item in the cost of weather services. In round figures we may take 130,000 feet (25 miles or 40 kilometres) as the practical limit of balloon sounding.

The radio-sonde is in essence an automatic radio transmitter broadcasting a signal which is modulated or characterized in some respect by the pressure, temperature, and humidity which are to be measured. To do this the meteorological measuring device, the sensor, must be coupled in some way to the circuitry of the radio transmitter and there are as many possibilities as there are variable factors in the arrangement. Generally speaking the sensor will be coupled to a resistor, capacitor, or inductor and the signal varied by amplitude or frequency modulation but there are many other arrangements including one in which the message from the transmitter is actually a morse code translation of the temperature or other reading.

The British Meteorological Office Mark 2B radio-sonde,

illustrated in Plate 14, is almost obsolescent after service for many years but later models are little more than refinements and the rather crude appearance of the assembly is an advantage for illustration. The three metallic shields attached externally to the radio transmitter contain the basic instruments, an aneroid barometer for pressure, a bimetallic thermometer, and a device of gold-beater's skin for humidity. In each case the change in the physical element produces a direct mechanical movement which is coupled to the armature of an inductor in an audio-frequency oscillator. Change in inductance in this case causes frequency modulation which may be measured by a cathode-ray tube at the receiver at the ground. Each element, for pressure, temperature, and humidity, is brought into action in turn through the action of the three-cup windmill driven by the airflow. An automatic receiver may be employed to pick up and record all the information.

This British radio-sonde weighs about $2\frac{1}{2}$ lb, costs about £12 at present, and can be carried by a balloon and hydrogen adding another few pounds to the cost. Perhaps a million of these instruments have been flown in different parts of the world over recent years and no doubt most of them now lie somewhere at the bottom of the oceans.

The radar wind-finding device consists of a simple metallic radar target carried by the balloon and followed by a standard radar at the ground. As the radar measures not only the angular position of the target but also the distance very accurately, the position of the balloon is known and the wind is computed from its displacement over short intervals of time in the horizontal direction. Half-minute readings from a balloon rising at about 1000 feet per minute give the detailed structure of the wind to the upper limit of the flight with an accuracy within a few miles per hour.

In the weather-radar, the equipment, which can indeed be the same as that used for measuring wind, takes advantage of the fact that water drops in the atmosphere act as reflectors of radar signals. As everyone knows, a radar scanning the vicinity of an airport may be used to pick up and follow the movement of every individual aircraft, the display being on a suitable radar screen. Naturally, radar will not show and follow each individual among the untold millions of raindrops but each region where rain is falling shows as a bright region on the display. The small droplets of cloudy air without rain or snow hardly affect the radar beam, so that a continuous radar record gives a rather faithful picture of the changing distribution of the rainy areas. The design of a radar suitable for weather scanning must of course take into account many factors. The wave-length of the radar beam must be such that the signal may penetrate to a useful distance giving a sufficiently powerful reflected signal throughout the range but not suffering too much attenuation. In Plate 13 is shown a Decca weather radar, installed at Singapore, working on a 3 centimetre wave-length with a peak power of 75 kW and giving a useful display of rain areas over a range of 100 miles or more. An example of a radar display is given in Plate 12. After a little experience with weather-radar one becomes very impressed by the beauty and power of the technique and is very surprised that – although it has been available for twenty years – it is not very widely used by meteorological services. Although a radar installation is not cheap, costing perhaps £10,000 (more or less according to the model), the expense of covering a country with a weather radar network is not excessive and we may expect more progress in this direction in the immediate future. There is certainly no technical reason, although there may be a financial one, why a forecast centre should not be almost completely informed of where rain or other precipitation is falling at all times and in great detail over a country-wide area. Unfortunately, as experience shows, the geographical patterns shown by regions where rain is falling change in structure rather quickly and even when a radar display is available it is not possible to predict the details for more than one or two hours ahead. Efforts have also been made for some years to adapt the radar to the quantitative measurement of rainfall amount over a large area as a substitute for numerous rain-gauges or as a very rapid indication of the risks of flooding but no great success has yet been achieved. The rate of rainfall depends on the number, size, and fall-speed of the raindrops, snowflakes, or hailstones, whereas the intensity of the radar echo depends on the population of drops in a different way, not at all on their fall-speed, and is different for water, ice, and wet ice. For these and other reasons there is no simple correspondence between rate of rainfall and radar echo at any instant of time, and even less correspondence between total rainfall over a period of time and any simple measure of radar response. If, however, radar is ill-adapted to the precise measurement of rainfall amount it may yet have an important future as the best available technique for estimating that most important hydrological factor, the total water that falls over a catchment area, at least in regions where for one reason or another adequate direct measurements by rain-gauge are impracticable – and most of the earth falls in this category.

Returning now to upper-air sounding we recall the effective limit to ballooning, which in round figures we took to be 130,000 feet, 25 miles or 40 kilometres and look for alternative load-carriers capable of breaking through this ceiling: there is the gun and there is the rocket. Let it be noted at once that while a couple of men with ordinary care and no great elaboration of equipment may fill and launch a balloon safely from any open space, the introduction of explosives and pyrotechnics brings all kinds of risks and difficulties. The administrative problems are comparatively formidable and the costs excessive unless the facilities of military installations and test ranges are put at the disposal of the scientist. Arrangements have, however, been satisfactorily made in a number of countries, the rocket having received almost all the attention to the exclusion of the gun, for various technical reasons which need not detain us.

The newest rocket to come on the market was designed to the specification of the British Meteorological Office and has been given the proper name 'Skua'. It is 8 feet in length, 5 inches in diameter, weighs over 80 lb and is capable of carrying a scientific load of 10 lb to a height of above 40 miles or 60 kilometres. The load is normally a radio-sonde which is released automatically to fall under the control of a parachute which acts also as a radar target. By this means useful temperature records are obtained, taking us through the very important warm layer around 50 kilometres described in the previous chapter. At the same time, tracking by a high-performance radar allows the winds to be measured. Rockets have been used regularly for meteorology for the past few years in North America while experimental rounds have been fired in other countries and it is not unlikely that a world-wide network will in time be established.

Much larger rockets attaining considerably greater heights have also been used for special purposes in a few countries, but the programmes of high atmosphere research take us far outside the study of weather and climate into the realm of space research which has deliberately been excluded from the scope of this book. We therefore end this chapter with a brief outline of the specifically meteorological space vehicle, the so-called weather satellite, so far an exclusively American initiative although the scientific benefits are available to all.

An earth satellite is a body which revolves in orbit around the earth and maintains its distance by a balance between the attractive force of gravity and – as we rather loosely say – the centrifugal force. The period of revolution to satisfy this condition increases steadily as the distance of the satellite is increased so that the moon, the earth's natural satellite at a distance of nearly a quarter of a million miles, takes over 29 days to complete its orbit. The artificial weather satellites are launched into an orbit quite near the earth, between 700 and 800 miles from the surface, which is sufficiently far into space to ensure little friction with the very tenuous atmosphere and a useful life of many months: the orbiting time at this height is little less than 2 hours.

It will be obvious that a high order of engineering skill, and corresponding engineering resources, are needed to build a rocket launching system capable of giving a satellite precisely the correct speed and direction of movement to ensure its entry into the desired orbit. More technical skill of a different kind is demanded in the design of the scientific instruments and radio equipment to be included in the satellite, and to make them sufficiently robust to withstand the explosive shock of the rocket-carrier. We shall, however, with the onlooker's insouciance, leave all these things to the technicians and briefly describe the artificial weather-satellite as though it had got into orbit in the same manner as did the moon – and no one knows for certain how and when that was.

The system, now under the control of the United States Weather Bureau, dates from the launching of *Tiros I* in January 1960 and is being continued with various modifications for an indefinite time: *Tiros IX* was launched from Cape Kennedy in January 1965. In principle there is an unlimited choice of heights and associated orbits. At one extreme the satellite might move round the earth from west to east, or east to west, and remain

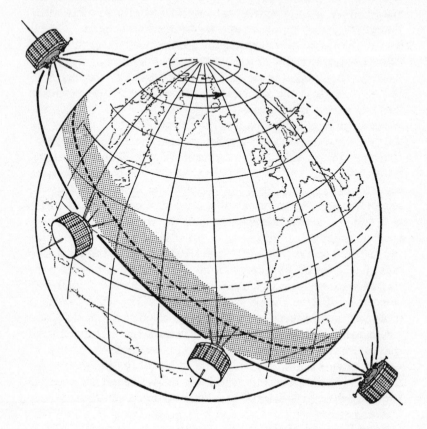

Fig. 3. Illustrating a weather satellite in orbit around the earth. The shaded area indicates the strip of earth which lies within the view of the satellite camera. As the earth twists about 28° in the time of one orbit, the consecutive strips are displaced about 28° towards the west. Only that half of the earth which is sunlit can be photographed by visible light. The use of infrared photography will permit the night time also to be covered.

always overhead at the equator, or it might even remain apparently stationary. At the other extreme it could be put into a polar orbit, that is one which passes overhead at each of the poles, but in practice an intermediate near-polar orbit is being used. With the help of Fig. 3 one may form the mental picture of the satellite moving round in space, held by the earth's gravity but unconcerned by the rotation of the earth once each day. In this way an eye looking down from the satellite will see below a strip of earth which is displaced by rather more than 28° of longitude towards the west on each successive circuit. After the course of a day and some thirteen circuits the satellite is more or less back on the initial orbit relative to the earth, having twice scanned the earth's surface – although half the time the orbit will be in shadow and the earth invisible unless the eye can see in the night. The eye is in fact a television camera which takes snapshots of the earth below, automatically at suitable intervals of time, stores the pictures electronically and transmits them for reception by suitable radio equipment within range on the ground. The design is such that each picture covers an area about 700 miles square, almost sufficient to include the whole of the British Isles, and by suitable timing a view of all the strip of earth below is recorded and transmitted. The meteorological interest is primarily in the distribution and types of clouds, and the results obtained have been of quite astounding scientific interest – far exceeding many expectations. The two illustrations given in Plate 15 will suffice to indicate the wealth of information which is yielded when the whole earth's surface is covered by such photographs.

The *Tiros* system is intended not only to provide data of scientific interest but also to serve operational needs, providing current data for the use of weather-forecasters. To this end the system now incorporates 'Automatic Picture Transmission', known as APT, whereby only a simple aerial and radio receiver coupled to a facsimile recorder are needed on the ground to receive, with very little delay, all the pictures transmitted by a satellite within range, which in practice is within a distance of over 1000 miles in any direction. The system is still in the experimental stage, but is expected soon to become fully operational and to continue at least as long as the United States continues to pay the bill in many tens of millions of dollars, while at the same time permitting anyone to tap the information with equipment costing perhaps a mere ten thousand dollars. The lack of sufficient information about clouds is not a major embarrassment for forecasters in the highly populated and advanced countries of Europe, so that their attitude to APT from weather satellites does not betray as much excitement as one might have expected with a technical innovation of such tremendous novelty. The reward will, however, be beyond argument for weather forecasts which relate to most other parts of the world, to the large areas of tropical lands, the southern hemisphere, and oceanic areas generally, for which weather data at present are far from adequate. Some striking successes have already been claimed in the early discovery of developing hurricanes.

The weather satellite has further valuable potentialities including: infrared photography which will extend its usefulness to the dark side of the earth, the measurement of sea temperatures, and the mapping of the limits of snow-fields and polar ice. There is every reason to hope that the system will eventually be established by international agreement as a permanent part of world weather observing. In providing an entirely novel method of observing the clouds, the satellite brings new life to the branch of weather-study which was the first to be systemized and which conveniently forms the subject of the next chapter.

Chapter 4

The Classification of Clouds

WATER evaporates from the earth's surface and is carried in the air as a completely transparent and invisible vapour; by some process, mainly by rising and expanding, the vapour-laden air is cooled to its saturation point and beyond, whereupon moisture is condensed in the visible form of clouds. When the cloud particles become large enough they fall to the earth as drizzle, rain, snow, sleet, or hail – precipitation to use the awkward but useful generic term – and so the cycle is complete. Put in this summary fashion we have a neat little story in atmospheric physics, attractive in its simplicity, both memorable and true, but as an account of the natural events to which it alludes it is about as complete and satisfying as would be a précis of Shakespeare's *Tempest* in a similar number of words. Perhaps not quite as many books have been written about clouds as about any play of Shakespeare's but, even so, they make an impressive collection on the library shelves and, furthermore, the problems remain both numerous and challenging. Almost any aspect of clouds, their size, height, composition, structure, electrification, colour, creation, or dissolution, raises questions to which only partial and uncertain answers can yet be given.

It would be difficult to overstress the importance of clouds as the necessary intermediary between invisible vapour and falling precipitation in the water cycle upon which all land-life depends, but their importance by no means ends here. Clouds which do not give rain, which never even threaten to give rain but which dissolve again into vapour before the precipitation stage is ever reached, have a profound effect on our climate. This is obvious

33

enough if we only think of the difference between a cloudy and a sunny day in summer or between an overcast and a clear frosty night in winter. Taking an overall average, about 50 per cent of the earth's surface is covered with cloud at any time whereas precipitation is falling over no more than say 3 per cent. Non-precipitating clouds are thus the common variety, rain clouds are the exception.

The climatic importance of clouds lies in their effectiveness in reflecting, absorbing, transmitting, and emitting radiation, to which further reference will be made in a later chapter. The effects are complicated because clouds are neither 'black' nor 'grey' but react to different parts of the spectrum quite differently. To the sun's visible radiation they are efficient reflectors, throwing up to as much as 80 per cent back to space, and so shining white in the eyes of the space traveller. What is not reflected mostly penetrates and is absorbed in clouds of sufficient vertical depth so that the amount of light reaching the earth is then quite small, as every photographer knows. Long-wave radiation from the earth, the invisible heat rays, is by contrast totally absorbed by quite a thin layer of clouds and, by the same token, the clouds themselves emit heat continuously according to their temperatures, almost as though they were black bodies. In this way clouds by day keep much of the sun's heat away, but at the same time and in the night-time too they return to the earth much of the heat that would have been lost. A completely cloudy day may be close and humid but never exceptionally hot, whereas during a cloudy night the temperature may hardly fall from its day-time value.

Clouds in the earth's atmosphere are virtually confined to the troposphere, that lower region extending to the base of the stratosphere at a varying height around 10 kilometres in middle latitudes, 17 kilometres in the tropics. It is true that occasionally, very occasionally, stratospheric clouds may be seen and, as scientific curiosities, they have attracted much interest, but they are sufficiently rare to be left out of account in the first general classification. It is a very good generalization to say that clouds are formed when the air becomes saturated by expansional (adiabatic) cooling, the expansion being the response to the reduction of pressure experienced when a parcel or layer of air moves up-wards. The confining of clouds to the troposphere is visible proof that this is the limit reached by air parcels from low levels: the overlying stratosphere is, in other words, the lid, but a fluid lid, to convection. The fact that the tropopause is nowhere a steady feature but is continually rising and falling, being folded or disrupted, or temporarily destroyed and reformed, is evidence that upward motions generated in the troposphere are turbulent in various ways and cause mixing with the stratosphere from time to time.

Of all the branches of meteorology, the study of the clouds in the sky must be given a special pre-eminence for not only do they present scientific problems of deep interest and subtlety but by their endless variety of form and structure, light and shade and colour, growth and decay and incessant movement as they are borne on the wind, they provide for anyone who will raise his eyes from the ground a mobile architecture which is no small part of natural beauty. Poets and painters have been moved and inspired to comment in their own way, but ordinary mortals may also capture the beauty in their cameras and the fascination of the minor art is witnessed time and again by the never-failing success of a discourse on clouds to any audience when the speaker goes armed with a selection of attractive slides; and if these have the excellence of modern colour transparencies a continual purring of applause will surely reward the lecturer however banal may be his comment. Almost every book on weather has its selection of cloud photographs to catch the eye of the reader, and the reviewer, and a number of delightful descriptive accounts have been published, profusely and beautifully illustrated. Quite one of the best, if not the most systematic and complete, is *Cloud Study* by F. H. Ludlam and R. S. Scorer, published by the Royal Meteorological Society, where the aesthetic interest of the illustrations is enhanced by expert comment, as the interest of a symphony might be enhanced by a musicologist. This is appreciation at its best, looking beyond the form to the organic structure and meaning of the clouds, but interest in clouds came long before scientific understanding. The accepted classification, long established and now formally agreed by international authority, is a descriptive classification deriving in a direct line from Luke Howard who, in 1803, was apparently the first to introduce the Latin words cirrus, cirrocumulus, cirrostratus, stratus, and cumulus. As it stands at present, and is fully described in the *International Cloud Atlas*, published by the World Meteorological Organization, the system recognizes ten genera with fourteen species, nine varieties and ten 'supplementary features and accessory clouds', all accorded Latin names: quite a little Linnaeus! The names of the genera are in regular and common use and unless one cares to speak, for example, of mares' tails, little sheep, jockeys, or cauliflower clouds there is no escape. The specific names are also essential and are therefore included in the following list of definitions copied from the *Atlas*. The photographs reproduced in Plates 1 to 11 illustrate most of the types.

Genera

CIRRUS	Detached clouds in the form of white, delicate filaments, or white or mostly white patches or narrow bands. These clouds have a fibrous (hair-like) appearance, or a silky sheen, or both.
CIRROCUMULUS	Thin, white patch, sheet, or layer of cloud without shading, composed of very small elements in the form of grains, ripples, etc., merged or separate, and more or less regularly arranged; most of the elements have an apparent width of less than one degree.
CIRROSTRATUS	Transparent, whitish cloud veil of fibrous (hair-like) or smooth appearance, totally or partly covering the sky, and generally producing halo phenomena.
ALTOCUMULUS	White or grey, or both white and grey, patch, sheet, or layer of cloud, generally with shading, composed of laminae, rounded masses, rolls, etc., which are sometimes partly fibrous or diffuse and which may or may not be merged; most of the regularly arranged small elements usually have an apparent width of between one and five degrees.
ALTOSTRATUS	Greyish or bluish cloud sheet or layer of striated, fibrous or uniform appearance,

totally or partly covering the sky, and having parts thin enough to reveal the sun at least vaguely, as through ground glass. Altostratus does not show halo phenomena.

NIMBOSTRATUS Grey cloud layer, often dark, the appearance of which is rendered diffuse by more or less continuously falling rain or snow, which in most cases reaches the ground. It is thick enough throughout to blot out the sun. Low, ragged clouds frequently occur below the layer, with which they may or may not merge.

STRATOCUMULUS Grey or whitish, or both grey and whitish, patch, sheet or layer of cloud which almost always has dark parts, composed of tessellations, rounded masses, rolls, etc., which are non-fibrous (except for virga) and which may or may not be merged; most of the regularly arranged small elements have an apparent width of more than five degrees.

STRATUS Generally grey cloud layer with a fairly uniform base, which may give drizzle, ice prisms or snow grains. When the sun is visible through the cloud, its outline is clearly discernible. Stratus does not produce halo phenomena except, possibly, at very low temperatures. Sometimes stratus appears in the form of ragged patches.

CUMULUS Detached clouds, generally dense and with sharp outlines, developing vertically in the form of rising mounds, domes or towers, of which the bulging upper part often resembles a cauliflower. The sunlit parts of these clouds are mostly brilliant white; their base is relatively dark and nearly horizontal. Sometimes Cumulus is ragged.

CUMULONIMBUS Heavy and dense cloud, with a considerable vertical extent, in the form of a mountain or huge towers. At least part of its upper portion is usually smooth, or fibrous or striated, and nearly always flattened; this part often spreads out in the shape of an anvil or vast plume. Under the base of this cloud which is often very dark, there are frequently low ragged clouds either merged with it or not, and precipitation sometimes in the form of virga.

Species

FIBRATUS   Detached clouds or a thin cloud veil, consisting of nearly straight or more or less irregularly curved filaments which do not terminate in hooks or tufts.

This term applies mainly to Cirrus and Cirrostratus.

UNCINUS   Cirrus often shaped like a comma, terminating at the top in a hook, or in a tuft the upper part of which is not in the form of a rounded protuberance.

SPISSATUS   Cirrus of sufficient optical thickness to appear greyish when viewed towards the sun.

CASTELLANUS   Clouds which present, in at least some portion of their upper part, cumuliform protuberances in the form of turrets which generally give the clouds a crenellated appearance. The turrets, some of which are taller than they are wide, are connected by a common base and seem to be arranged in lines. The castellanus character is especially evident when the clouds are seen from the side.

This term applies to Cirrus, Cirrocumulus, Altocumulus, and Stratocumulus.

FLOCCUS   A species in which each cloud unit is a small tuft with a cumuliform appearance, the lower part of which is more or less ragged and often accompanied by virga.

This term applies to Cirrus, Cirrocumulus, and Altocumulus.

STRATIFORMIS   Clouds spread out in an extensive horizontal sheet or layer.

This term applies to Altocumulus, Stratocumulus, and, occasionally, to Cirrocumulus.

NEBULOSUS   A cloud like a nebulous veil or layer, showing no distinct details.

This term applies mainly to Cirrostratus and Stratus.

LENTICULARIS   Clouds having the shape of lenses or almonds, often very elongated and usually with well-defined outlines; they occasionally show irisation. Such clouds appear most often in cloud formations of orographic origin, but may also occur in regions without marked orography.

This term applies mainly to Cirrocumulus, Altocumulus, and Stratocumulus.

FRACTUS   Clouds in the form of irregular shreds, which have a clearly ragged appearance.

This term applies only to Stratus and Cumulus.

HUMILIS   Cumulus clouds of only a slight vertical extent; they generally appear flattened.

MEDIOCRIS   Cumulus clouds of moderate vertical extent, the tops of which show fairly small protuberances.

CONGESTUS   Cumulus clouds which are markedly sprouting and are often of great vertical extent; their bulging upper part frequently resembles a cauliflower.

CALVUS   Cumulonimbus in which at least some protuberances of the upper part are beginning to lose their cumuliform outlines but in which no cirriform parts can be distinguished. Protuberances and sproutings tend to form a whitish mass, with more or less vertical striations.

CAPILLATUS   Cumulonimbus characterized by the presence, mostly in its upper portion, of distinct

39

cirriform parts of clearly fibrous or striated structure, frequently having the form of an anvil, a plume or a vast, more or less disorderly mass of hair. Cumulonimbus capillatus is usually accompanied by a shower, or by a thunderstorm, often with squalls and sometimes with hail; it frequently produces very well defined virga.

The following names of supplementary features and accessory clouds are also frequently met:

INCUS The upper portion of a Cumulonimbus spread out in the shape of an anvil with a smooth, fibrous, or striated appearance.

MAMMA Hanging protuberances, like udders, on the under surface of a cloud.
This supplementary feature occurs mostly with Cirrus, Cirrocumulus, Altocumulus, Altostratus, Stratocumulus, and Cumulonimbus.

VIRGA Vertical or inclined trails of precipitation (fallstreaks) attached to the under surface of a cloud, which do not reach the earth's surface.
This supplementary feature occurs mostly with Cirrocumulus, Altocumulus, Altostratus, Nimbostratus, Stratocumulus, Cumulus and Cumulonimbus.

PILEUS An accessory cloud of small horizontal extent, in the form of a cap or hood above, the top or attached to the upper part of a cumuliform cloud which often penetrates it. Several pileus may fairly often be observed in superposition. Pileus occurs principally with Cumulus and Cumulonimbus.

The main virtue of the descriptive classification for practical application is its objectivity and independence of theory, permitting an observer at an isolated station with only a modest scientific training to give a name to any cloud he sees. He does not need to decide for himself upon the mechanism by which the cloud was formed or maintained, often a matter of opinion difficult to be confident about with no evidence other than appearance to go upon, and even a matter of controversy among experts, for only in very recent years have plausible explanations been given for some of the peculiarities. But the scientific interest lies in the uncertainty and controversy, not in the Latin name, and it is more interesting to discuss why the clouds appear as they do on any occasion than to hang on to them the agreed label.

It is not quite true to say that all clouds are formed by cooling occasioned by expansion and we may as well dispose of the main exceptions first. It is true that the air must become saturated, or at least very nearly saturated before cloud particles will form and this means either the addition of water vapour or the reduction of temperature. It is not easy in nature to cause condensation by adding pure water vapour to the air, for the simple reason that evaporation naturally stops before condensation begins, but in special circumstances something like the cloud from a steaming kettle does occur. A localized example is the cloud over a hot spring or geyser; more widespread is the mist rising from a stream in frosty weather or from relatively warm ocean waters when very cold winds blow across them; more trivial is the man-made cloud from the exhaust of a motor-car in frosty weather, especially when the engine is cold; more remarkable is the man-made cirrus cloud from the exhaust of a high-flying aircraft, the pure white rippled condensation trails which may stretch across the whole vault of the sky, strikingly beautiful for a few minutes before either dissolving away or dispersing into 'ordinary' cirrus, revealing their origin thereafter only to the discerning eye. In all these cases the essential process has been the addition of water vapour to the air but in every case also the detailed physics is rather complicated, for heat as well as moisture is added and it is a matter for precise calculation if theory is to decide whether in the outcome the saturation point is ever reached. The exhaust from a well-tuned car is generally invisible unless the outside air is either cold or unusually moist and near saturation already. The rather subtle nature of the problem is realized by noting that very near the exhaust outlet there is clear air, for here the temperature is too high for saturation; then there is a plume of cloud bearing witness to the fact that in the mixing process the effect of cooling surpasses that of dilution, so permitting saturation to be surpassed for a time; but soon afterwards the plume disappears once more by further mixing and evaporation, showing that now the effect of dilution of moisture by the surrounding air surpasses the effect of further cooling, with the consequence that the air becomes unsaturated once more. The case of the aircraft exhaust is similar but complicated by temperatures well below freezing-point and the possibility of forming ice clouds. It is, however, a problem which is amenable to calculation in a rather interesting way. The fuel is a known composition of hydrocarbons which is burnt almost completely so that the rate of production of water vapour, H_2O, is decided by the rate of fuel consumption. All the energy of combustion goes either in heating the exhaust gases or in driving the aircraft and there are rather reliable design data which allow an estimate to be made. In this way the amounts of exhaust heat and of exhaust moisture may both be calculated and further calculations will show whether by mixing this hot moist air with the cold atmosphere a sufficient degree of saturation will be reached. It is an interesting calculation which leads to the rather surprising result that persistent trails cannot occur unless the air temperature is a long way below freezing-point. The point was of concern to the weather-forecaster at one time as being of some military importance, long condensation trails being far more easy to detect than the high-flying aircraft itself, and methods of avoiding trail formation by using special fuels and in other ways were seriously studied. For peaceful purposes the phenomenon is of little moment except perhaps as a reminder that human activities are capable of altering the higher atmosphere in a detectable way and possibly in a significant way if that were the deliberate aim.

Direct cooling as a method of cloud formation is important for cloud lying on the ground, that is fog, but in the free atmosphere direct cooling, not occasioned by expansion, is usually too slow a process to be the critical final cause. One should not, however, overlook the important process whereby maritime tropical air, which brings so much low cloud and drizzle, is brought near saturation by cooling during its long ocean passage from lower latitudes. The effect of direct cooling may also become paramount in the maintenance of a layer of cloud once formed, for the cloud top cools by radiation. As in so many other meteorological problems, attractively simple generalizations on cloud formation must, in the respect for truth, be hedged about with numerous exceptions and saving clauses but it remains substantially true that clouds are formed by expansion and adiabatic cooling and that, in consequence, a dynamical classification is provided by a classification of the various modes and scales of upward motion of the atmosphere. Dynamical classifications of clouds on this basis have been given in the textbooks for many years, undergoing improvements and modifications as our understanding of the dynamics and microphysics has progressed, and the time is ripe for the formulation of a comprehensive up-to-date system of which Table 1 may be regarded as a skeleton.

TABLE I.   CLASSIFICATION OF PROCESSES OF CLOUD FORMATION

	Description of cloud
1. Addition of moisture	
(*a*) At the earth's surface	Steaming fog
(*b*) In the free atmosphere	Condensation trails
2. Cooling at the earth's surface by conduction and radiation	Fog

Cooling by ascent and expansion

3. Mechanical lifting over high ground	Fog on high ground or layer cloud at higher levels (special names: banner cloud, föhn cloud)
4. Wave motion set up by topography	Lenticular clouds at any height in the troposphere
5. Turbulence	
(*a*) Turbulence originating at the earth's surface	Stratus or low Stratocumulus sheets
(*b*) Turbulence originating in the free atmosphere	Stratocumulus or Altocumulus sheets
6. Vertical convection	
(*a*) originating at the earth's surface	Cumulus or Cumulonimbus
(*b*) originating in the free atmosphere	Castellanus or Cumulonimbus

7. Slow ascent over large areas
(mainly in depressions)

(*a*) Frontal	Clouds in layers: Cirrostratus, Altostratus, Nimbostratus, sometimes without clear lanes
(*b*) Non-frontal	Layer clouds at any level

Note. Many clouds are produced by a combination of more than one process, e.g. turbulence plus convection, frontal ascent plus convection, etc. Cloud forms are also modified by additional processes, e.g. falling precipitation, subsidence, downward instability (mamma).

The elaboration of cloud study on the basis of the dynamical classification is a large part of the science of meteorology and will naturally arise in the course of later chapters of this book concerned with vertical convection, synoptic weather systems, and the general circulation of the atmosphere. It is convenient therefore to leave the subject at this stage and to turn attention to the study of clouds from a totally different point of view.

Chapter 5

The Microphysics of Clouds

THAT part of weather study which we call the microphysics of clouds, is concerned with the analysis of the processes within the clouds, pursued even to the behaviour of the individual particles themselves. It is a study of quite remarkable events, unnoticed by the casual observer and hardly dreamt of until awkward questions are asked and explanations are sought, and it is therefore illuminating to open by asking some of these questions.

When invisible water vapour in the air reaches saturation point, or sufficiently close to saturation point, enormous numbers of liquid droplets (or solid ice particles) are produced, becoming visible as a 'cloud'. What determines the degree of saturation or supersaturation necessary to produce cloud? What determines the number and size of the cloud particles? Under what circumstances does condensation take place directly into ice particles? Under what conditions do water droplets freeze or do ice particles melt? What happens to a mixed population of water droplets and ice particles? What factors decide whether a 'cloud' of small floating particles will give rise to 'precipitation' of larger falling particles? What are the processes involved in the transition from cloud to precipitation? What are the special conditions distinguishing the various forms of precipitation, drizzle, snow, sleet, and hail from ordinary rain?

Searching for answers to these questions one is brought up against the need for nuclei of condensation and a range of new questions: what is the chemical composition and physical state of the nuclei; what is their origin; how are they transported and eventually removed from the air; are special nuclei required for

45

the formation of ice particles? Looking more closely at the ice phase many notably different types of single crystals and agglomerations are observed, setting in train a new set of inquiries into why, how, when, and where, and still the range of problems is far from exhausted. Many of the curious and intriguing optical phenomena in the atmosphere are caused by ice or water particles, rainbows, glories, coronae and haloes of many kinds, while no one can for very long fail to ask how it is that a cloud sometimes manages to produce the electric charges, and discharges, necessary for that most violent and spectacular display of nature – the thunderstorm. It is sometimes said that scientific research inevitably raises more problems than it solves and no better support for the thesis could be wished for than the case of cloud microphysics. The subject at the present time is expanding and developing more quickly than ever before, and if we can now give acceptable answers to many questions with far more certainty than we could twenty years ago there are still innumerable doubts and difficulties and the science is in a healthy state.

Fundamentally, the problem is that of the changes of state of water substance, H_2O, between its vapour, liquid, and solid states, a field of classical physics in which some very firm answers can be given. The production of visible cloud is the condensation from vapour to liquid or solid and we may as well begin at this stage. We generally say that the air can hold no more than a definite maximum amount of invisible gaseous water, more or less according as the temperature is high or low, but the statement is acceptable only with reservations. In the first place, the presence of air – that is the unpolluted mixture of pure permanent gases – has little to do with the process. It is the amount of vapour in the available space that matters, the number of molecules of H_2O per cubic centimetre, and the presence of the other gases is not directly relevant. In this respect, it might be more correct to say that the space and not the air is more or less saturated with vapour, but in meteorology insistence on this distinction would be quite unnatural and confusing. The air is always present, the vapour contributing rarely more that 1 per cent by weight and usually very much less than this; the vapour moves with the air, has the same temperature as the air and expands or contracts with the air. For most purposes it is helpful to think of the vapour as being in the air and quite unambiguous to say that the air is dry or moist, has a low relative humidity or is saturated. However, in considering the physics of phase change the presence of the air is not always relevant.

When liquid water and gaseous vapour are present side by side one needs only to think of the exchange of molecules across the interface to have a clear mental image of evaporation and condensation going on continuously. The molecules in the liquid are in incessant motion and a small proportion, moving more rapidly than the average, escape from the liquid surface by overcoming the inter-molecular attractive force which binds the liquid together: in much the same way a rocket, given sufficient speed, will escape from the earth's gravitational force. The warmer the liquid the greater the speed of the molecules and the greater the number which have the necessary escape velocity – the warmer the water the more rapid the evaporation. At the same time, any molecules from the vapour which penetrate the liquid surface are captured and condensation takes place, at a rate which depends on the vapour temperature and density – or the vapour pressure. The net effect of the two processes going on continuously is either condensation or evaporation and there is a state of balance when escape and capture are at the same rate: in this case, the air is just saturated with respect to the liquid surface.

It has been necessary to labour over this image of the processes in terms of molecular movements in order to appreciate the difficulties which arise when the vapour exists in the atmosphere far removed from any liquid surface. The air might be supersaturated, in the sense that if there were liquid present the vapour would quickly be captured by it, but in the absence of any liquid there is no obvious reason why condensation should ever begin and experiment proves that the argument is a valid one. If air is carefully purified by filtering, it will not produce cloud droplets even if cooled by expansion far beyond its normal saturation point or dew point. C.T.R. Wilson, working with his famous expansion cloud chamber, was able to show this quite conclusively late in the last century. His method of purifying the air was to allow the droplets produced during cloud formation to settle out of the chamber and to repeat the process several times with the same sample of air. Ultimately four-fold supersaturations, that is humidities of 400 per cent, were necessary to produce condensation in the purified air.

These results, obtained first by Wilson and broadly confirmed by many later experimenters, have a very important bearing on natural meteorology, not because supersaturation occurs in the atmosphere but because it does not occur: why is it that in the atmosphere condensation to clouds invariably happens as soon as normal saturation is reached? The answer is that the natural atmosphere, however clean it may appear to be, is always supplied with a sufficient number of minute particles of salts, acids or other substances which serve just as well as liquid water in capturing water molecules from the vapour. These are the 'nuclei of condensation', and are effective as soon as the air becomes even slightly supersaturated. As a matter of fact, there are many observations of clouds in air whose relative humidity is considerably below 100 per cent, evidence of nuclei which are hygroscopic, but methods of measurement within natural cloud are not sufficiently refined to prove that even slight supersaturation ever occurs. If for practical purposes we assume that cloud will always form in the atmosphere when ordinary saturation is attained (that is relative to a flat surface of pure water), we shall not go far wrong. Microscopic counting shows that the droplets forming in an ordinary cloud are measured in hundreds or even thousands to the cubic centimetre, millions to the litre, numbers which may strike one as incredibly large until we become familiar with the minuteness of the particles with which cloud physics has to deal. We shall need to return to these problems of nuclei and droplets when the process of raindrop formation is considered, but meanwhile we may note that although the nuclei are extremely numerous they are generally quite invisible and utterly negligible as a contribution to the weight of the air, which can still be treated as a 'perfect gas' in most meteorological calculations. Some representative numbers are given in Table 2.

When the air contains exceptionally large amounts of dust or smoke products it is noticeably hazy but the cleanest of air on days of excellent visibility is yet well laden with the finer nuclei of condensation. Cloud droplets measuring on average say 10μ* (0·01 millimetre) in diameter and occurring in numbers of 1000

* μ = micron.

TABLE 2.   SOME ORDERS OF MAGNITUDE IN CLOUD PHYSICS

Particle	Diameter in centimetres	Mass grammes	In 1 cubic metre of air Number	In 1 cubic metre of air Mass grammes
Air molecule	10^{-8}	10^{-22}	10^{25}	1000
Small nuclei of condensation	10^{-5}	10^{-15}	10^{9}	10^{-6}
Giant nuclei of condensation	10^{-3}	10^{-9}	10^{3}	10^{-6}
Cloud droplet	10^{-3}	10^{-9}	$10^{8}, 10^{9}$	$0 \cdot 1, \quad 1$
Drizzle droplet	10^{-2}	10^{-6}	10^{6}	1
Raindrop	10^{-1}	10^{-3}	10^{3}	1

per cubic centimetre are spaced at about 1 millimetre apart and less than 1-millionth of the whole volume consists of water. But such an aggregate is quite different from the aggregate of nuclei and is very much a visible cloud. A thickness of about 100 metres will suffice for almost every ray of light entering the cloud to strike a droplet many times and be absorbed, reflected, or refracted, so casting a strong shadow in sunlight and destroying the definition of any object looked at through the cloud, that is producing bad visibility – indeed a technical 'fog' to an observer within the cloud.

The nature and formation of clouds of ice particles may be considered next because they are remarkable in a number of ways, and far more important than might reasonably have been expected. The temperature in the atmosphere is often below that of the freezing-point of water, even near the earth's surface, and it is rarely necessary to climb beyond 15,000 feet – say 5 kilometres – to find freezing temperatures even in the tropics, so that clouds in the atmosphere are very often at sub-freezing temperatures and ice clouds are commonplace. Presented with these facts for the first time, even a trained physicist caught off his guard might be excused for supposing them to be so commonplace as to raise no particularly significant problem. If the temperature is below freezing-point the particles will freeze and if it rises again above freezing-point they will melt but otherwise there is no essential difference, or so it might quite erroneously be imagined. The first fallacy is to assume that water naturally freezes at temperatures below the freezing-point, an assumption which is so unfounded as to cause some punctilious physicists to refuse to use the term 'freezing-point' for 0°C or 32°F. It is true that ice becomes unstable and passes spontaneously into water if it is supplied with heat at this critical temperature: pure ice cannot be made warmer that 0°C and therefore this may properly be called the 'melting-point' of ice, but pure water may, in certain circumstances, be cooled far below 'freezing-point' to become 'supercooled'. The phenomenon has much in common with the supersaturation of the vapour for, just as the widely separated particles of a vapour cannot by chance get together in sufficient numbers to initiate the liquid phase, so the mobile liquid molecules cannot by chance settle into the crystalline form of a solid although this, when attained, is the more stable state at sub-freezing temperatures: once more a special nucleus is needed. Not all the details of super-cooling are yet understood but it appears that a special nucleus is necessary for all freezing until the temperature falls to about −40°C when all droplets freeze, and that such special nuclei are present in the atmosphere in very varying numbers. Moreover, no nucleus, other than ice itself, is effective until the temperature falls several degrees below 'freezing-point', so that supercooling in the at-mosphere, far from being exceptional, is quite normal wherever the temperature is suitable. Furthermore, to add another fact of critical importance, freezing nuclei are generally very much less numerous than condensation nuclei or than water cloud droplets with the result that ice crystal clouds have a low particle count, and when a water cloud does begin to freeze it does so only in a small proportion of particles resulting in a mixed cloud, with interesting further consequences.

The regular occurrence of supercooling is directly responsible for a number of rather curious facts of meteorology. Cold-weather fogs are often at sub-freezing temperatures and are then almost always supercooled water droplet fogs which readily freeze on impact with any solid body. They are responsible for the annoying and dangerous icing of motor-car windscreens and also – surely as adequate compensation even to the motorist – for the strikingly beautiful displays of rime which sometimes clothe every hedge and tree on a sunny winter's morning after a night of frost and fog. But what is a minor hazard to the motorist has been a major problem in the history of aviation ever since the days, some thirty years ago, when it was becoming common practice to fly through clouds on instruments instead of avoiding them, and the ultimate target of routine all-weather flying, only just being reached today, was already within the sights of far-seeing aircraft designers and airline planners. It was discovered, not without tragic incidents, that supercooled water was liable to be encountered at any flying level and in many types of clouds in varying degree and with varying importance. Particularly common was the accretion on leading edges of wings upsetting the airflow, increasing the drag and reducing the lift; or heavy accretion on airscrews leading to alarming vibrations and even physical damage to other parts occasioned by pieces of ice which had broken away. The running battle against icing has been continuous but with the outcome never seriously in doubt thanks to the combination of meteorological understanding and training, wise air-pilotage, the general introduction of de-icing equipment and the ever advancing performance of modern aircraft. However, even the recent introduction of jet engines led to a new crop of unexpected trouble when it was discovered that in certain conditions ice could accumulate in air intakes and be followed by sudden engine-failure.

The occurrence of supercooling has then presented man with an interesting batch of curious facts and special problems but, apart from these, the phenomenon has a bearing on the production of rainfall and other forms of precipitation much more far-reaching than one might expect to find. It is now opportune to look into this matter, and to note that once again the naïve assumption is far from the truth.

Rain, we readily agree, is produced by the condensation of the excess water vapour in the air when this is cooled by ascent and expansion: the condensed moisture forms clouds of little drops which in time grow big enough to fall to the ground by gravity. All very well at first sight, but a little further examination brings one up against the facts of the fall-speed of droplets in air. The weight of a sphere depends on its volume or the cube of its radius, whereas the air resistance depends in rather a complicated way on its fall-speed and radius. The result is that for a large and heavy body, air resistance is a secondary factor and it falls 'like a stone', but for a small body, such as a cloud droplet, the air resistance is all-important and it falls 'like a feather' only at its 'terminal velocity' – the speed at which the air resistance just balances the force of gravity. These speeds have been measured experimentally for drops of different sizes, and a few figures are included in Table 3. These values do not follow the theory for falling spheres because the larger drops are distorted by the air-flow. Drops larger than about 3000μ (3 millimetres radius) disrupt, with the interesting result that rain can never fall through the air more quickly than about 10 metres per second.

TABLE 3.   FALL-SPEEDS OF WATER DROPS IN AIR
(0°C, 900 mb*)

Nature	Diameter in Centimetres	Fall-speed centimetres per second	Time to fall 1000 metres
Cloud	0·001	0·3	5 days
	0·002	1·3	1 day
	0·004	5·4	6 hours
Drizzle	0·01	27	1·0 hours
	0·02	76	21 minutes
	0·04	170	10 minutes
Rain	0·1	390	4 minutes
	0·2	690	2·3 minutes
	0·4	930	1·8 minutes

* mb = millibar.

From the table we note that cloud droplets averaging about 10μ in radius would have no time to reach the ground from middle-cloud levels even in the whole life of a depression. The small size is due to the ubiquitous condensation nuclei of which some 1000 per cubic centimetre may become active, and the excess water distributed among so many drops permits only the small size. 1000 drops of 10μ radius in each cubic centimetre account for 4 grammes of water per kilogram of air, a very considerable figure. The droplet radius and so the fall-speed is, moreover, not very sensitive either to the number of nuclei or to the amount of condensed water. Thus by increasing the total water content or decreasing the number of droplets eight-fold, the diameter is only doubled and the fall-speed increased four-fold; the drops are still very small and not at all like raindrops. The firm conclusion is that if there were no process for collecting the condensed water together into a relatively few really big drops, each the equivalent of perhaps a million cloud droplets, the clouds could remain in suspension almost indefinitely and rainfall would be very slight. The effect on world climate would be quite catastrophic. The atmosphere would become nearly saturated everywhere and the air generally cloudy; there would be little precipitation and of course very little evaporation and very little sunshine; what the consequences would be for life on Earth is hardly worth speculating upon, but they would be profound. Often it seems that matters which control our very existence are the outcome of side-effects in the laws of physics which may be looked upon either as accidents or as special dispensations of providence according to our point of view: that there is a natural process capable of changing clouds with a million droplets to the litre into rain with only one drop to the litre, a tremendous transformation, is a circumstance of this kind. What the process or processes responsible for the metamorphosis may be, has been among the more hotly debated of meteorological problems.

For a time experts were almost persuaded that to overcome one curious and stubborn property of nature, the almost colloidal stability of cloudy air, it was necessary to invoke a special dispensation, namely the fact that the temperatures in the upper atmosphere are generally well below freezing-point and that ice also enters the picture. This was as much as to say that we cannot explain the very common phenomenon of the transition from water vapour to rainfall without introducing the freezing process. In spite of an ingrained reluctance to explain a common event as the result of some otherwise unrelated factor – that is as a fluke of nature – the argument due to Bergeron (1933) and strongly supported by Findeisen (1938) was in this case persuasive. It goes something like the following. At a temperature some degrees below freezing-point, say minus 5–10°C, a relatively small number of freezing nuclei come into action and perhaps one water droplet in a million freezes. At once the stability of the cloud is changed, for at sub-freezing temperatures water vapour condenses much more readily on an ice crystal than it does upon a water drop, or, perhaps more accurately, water molecules evaporate from water droplets much more freely than they do from ice crystals. The result is that in a mixed population of droplets and ice particles in an environment of vapour the droplets dwindle and the ice particles grow until in due time the droplets have disappeared and the ice has captured all the excess vapour. Clearly if this qualitative argument is to be upheld time is of the essence and careful calculations do show that the process can account for the complete transformation of a supercooled cloud in a matter of minutes. A vital point in the theory is of course that the ice cloud shall be composed of a relatively small number of large particles comparable in size with raindrops and so capable of falling rapidly to earth, melting to raindrops in the warmer lower atmosphere. Putting this theory in a nutshell, we must say that all rain starts as snow, and there are a number of facts which make this astounding assertion rather acceptable, at least to meteorologists of middle and higher geographical latitudes. Rain clouds are almost always thick enough in the vertical to reach well beyond the freezing level so that the process is at least permitted by the temperatures. Furthermore, mountain climbing and, later, aviation have made us familiar with the fact that when it is raining at low levels it is often snowing higher up. And the observation of large cumulonimbus clouds, the clouds of showers and thunderstorms which often have the virtue of being visible as a whole and available for inspection from top to bottom, have long been known to undergo a rapid transformation at the time when rain is about to begin. The upper levels, in the anvil, become fibrous and less opaque as would be the result of 'glaciation' and the whole depth of cloud, previously rainless, suddenly, in a matter of minutes, may become a wall of falling snow and rain. There is indeed quite enough evidence to show that the freezing process is often an important feature in the metamorphosis of cloud to rain, but to say that it is essential is to go too far as was slowly conceded by the advocates of the theory.

The first and most convincing evidence that rain could fall from non-freezing clouds was its frequent occurrence in low latitudes in circumstances where the cloud itself nowhere reached sufficiently high to meet temperatures below the freezing-point. The fact was never in doubt amongst those experienced in tropical weather, and the rather curious persistence of the belief in the freezing process as a necessary condition can be accounted for only by the lack of an adequate interchange of ideas, the difficulty of verifying the observations, and the limited amount of documented evidence available in the published literature. But gradually, in the course of ten years or so, all doubts were removed and the occurrence of rainfall produced by the coalescence of cloud particles in 'warm clouds' was generally accepted, setting in train a long course of theoretical, experimental, and observational research required to clarify the process and to determine the relative importance of coalescence and freezing in the initiation and development of rain in different circumstances. In order to appreciate and to overcome the difficulties confronting the explanation of the formation of raindrops without invoking the aid of ice particles, it is necessary to return to the process of cloud formation and to go into matters in more detail than we have done so far. When the cloudless air ascends, for any reason, it cools by expansion at the definite rate of $1°C$ for each 100 metres of ascent ($5.4°F$ each 1000 feet) and condensation begins as soon as the saturation point is reached. Thereafter, allowance being made for the liberation of latent heat, the cooling rate in further ascent is considerably less by an amount which depends upon the temperature. This rate is precisely given by theory, an average value being about half the dry adiabatic value, say $0.5°C$ per 100 metres. Evidently then, the ascent needs to continue a considerable distance before appreciable liquid moisture is made available for condensation into droplets but, if we neglect mixing with drier air and fall-out, the amount of liquid water increases steadily with the height attained as the amount of vapour correspondingly decreases. In order to fix ideas we may take as an example the case of condensation at a cloud-base temperature of $10°C$ ($50°F$) and compute the liquid water content at different heights in the cloud to obtain the following values:

Height in cloud	Temperature	Mass of Water per kg of air	
		Vapour	Liquid
Base	10°C	8·7 g	0
1 km	5°C	6·9 g	1·8 g
2 km	–1°C	5·2 g	3·5 g
3 km	–7°C	3·6 g	5·1 g

Assuming 1000 droplets per cm³, the mean-volume radius of droplets is 8μ, 10μ, 12μ at 1, 2, 3 km respectively and the fall-speeds in still air are 0·9, 1·3, and 2 centimetres per second, not a large range. The mixing within the cloud might be supposed to produce a mixture of sizes with relative falling speeds of the order 1 centimetre per second and an upper limit to the rate of growth by collision might be estimated by supposing that a droplet of radius say 12μ were to fall at 1 centimetre per second and sweep up all the free water in its track. We find that the drop of mass 7×10^{-9}g would gain mass at the rate of 5×10^{-12} g/sec with a cloud water content of 1 g/m³, or at 5×10^{-11} g/sec with a cloud water content of 10 g/m³, requiring therefore in the two cases 25 minutes or 2·5 minutes to double its mass. The rate of growth would, however, increase rapidly with the size of the droplet, partly because the droplet is larger and partly because it falls more rapidly, and the combined effect would cause a droplet to grow roughly exponentially, that is to double its mass in an equal time whatever its size. Taking the first case of cloud of 1 g/m³, the doubling time is 25 minutes and for a 12μ droplet to become a 120μ drizzle droplet, an increase of mass 1000-fold, it would need to double about 10 times, requiring about 4 hours. With 10 g/m³ the drizzle droplet could, however, grow in about 25 minutes. The growth from drizzle size to large droplet size, a thousand times heavier still, could be accomplished in a further period of about the same length provided there were sufficient depth of cloud to fall through. Naturally this rough calculation is no more than an indication of the speed of the process but it does explain why it is that clouds need to be quite thick, at least 1 km, before they have a chance to produce even drizzle drops, and why it is that even deep convection cloud must persist for about half-an-hour if rain is to fall. The main criticism of the argument lies in the fact that very small droplets have a marked tendency to avoid collision with other droplets, being swept aside by the air motion, and it is believed that the collision process cannot become efficient until sufficient droplets grow to more than 20μ in radius. How this is accomplished is not easy to say, but the most plausible hypothesis is that there are present in the atmosphere a small proportion of 'giant nuclei' which do grow droplets quickly by direct condensation to this larger size, and actual measurement of cloud droplet sizes does confirm that the larger size is generally present. A number of careful calculations have been carried out by various cloud physicists, and Professor B. J. Mason, for example, concluded a few years ago that, 'it seems entirely reasonable that drops of radius 350μ should develop in shallow maritime clouds within periods of about one hour', and went on to explain what has often been remarked upon – how maritime clouds may precipitate more readily than do clouds over land. In clean air over the sea the total number of condensation nuclei is low and the number of cloud droplets may be less than $100/cm^3$ whereas in polluted air the numbers may be ten or more times larger and the average sizes for a similar cloud that much smaller. The larger droplets, it is argued, would be caught in collisions more efficiently than the smaller, although if the theory is sound, the supply of giant nuclei is also important and little is known on this point.

It has proved to be no easy task to determine the nature of the atmospheric nuclei which are effective in the initiation of condensation in natural cloud formation, but the accumulation of evidence by many investigators using a number of quite different methods is that numerous components of the aerosol are effective in different circumstances. We may, by a little reflection, think of a number of processes which evidently introduce particulate matter into the atmosphere and which may therefore be possible sources of nuclei. There is spray from the sea which, on evaporation, will leave particles or concentrated droplets of sea salts; there is the mechanical raising of dusts by the wind, a process which must seem very promising to anyone who has experienced the dust-laden air carried in the wind hundreds of miles from a desert source; there are the products of combustion, from artificial man-made fires polluting the air continuously and from natural forest and heath fires, a by no means negligible process; there is the output from active volcanoes, perhaps a minor contribution for most of the time but capable of visibly polluting the upper atmosphere on a world-wide scale on the rare occasions of major new eruptions; and finally, there is the extra-terrestrial source, cosmic dust entering the atmosphere continuously as meteorites, a source which recent research has brought into fresh prominence.

Bearing in mind that the earth is largely ocean-covered and that breaking waves are a commonplace wherever the winds are more than fresh, sea salt might seem the most likely source and certainly, being hygroscopic, it provides an efficient nucleus of condensation. But as G.C. Simpson pointed out, if rain is formed by the combination of cloud droplets the number of nuclei returned to the earth in rain is so large as to make it difficult to believe that the mechanical disruption of spray could be capable of producing them. The average rainfall of the earth is not far from 100 centimetres per year or 100 cm^3 of water over each square centimetre of the surface, and if this amount of water is obtained by the combination of droplets of radius 10μ, each containing a nucleus, the number of nuclei returned to the earth is some $2 \cdot 5 \times 10^{10}$ /cm^2 per year, or 1000 /cm^2 per second. Simpson, using rather different figures, estimated that to perform the task the rate of production where waves were breaking would need to be some 50,000 /cm^2 per second which seemed virtually impossible. It is, however, difficult to be at all sure of what may be the outcome of natural processes on the microscopic scale and Mason, on the basis of ingenious laboratory experiments, was led to the conclusion that the bursting of minute bubbles arising from the sea surface could in fact plausibly produce 1000 nuclei /cm^2 per second which is precisely the number required by our elementary estimate. It is therefore not at all unreasonable to believe that, were there no other source of nuclei, sea salt would be sufficient to produce clouds and rain much in the manner observed and so little is known of the nature of the aerosol over the open oceans that it would be unwise to conclude that sea salt is not in fact the effective agent for most of the rain that falls on the earth's surface.

Other evidence does however show beyond doubt that many other nuclei exist and are active, perhaps the most convincing evidence being that obtained by studying under the electron microscope the actual residues left after evaporating collected droplets. The work is not easy and few results have been published, and if there is any lesson to be learnt from aerosol analyses it is that its composition is extremely varied, making it dangerous to generalize. Yet there is no reason to disagree with Professor Mason's summing up, provided we remember the limited evidence from the open oceans, that 'of the nuclei involved in cloud-droplet formation, perhaps one-tenth consist of sea salt, and the rest of mixed nuclei and the products of natural or man-made fires'. It could, however, be the case that the very large numbers of nuclei which form cloud droplets are not so important as the minority of giant nuclei which form the larger droplets and account for the bulk of the water. In this case nuclei formed by the condensation of combustion products, probably for the most part very small, may do little to help the deposition of rain, and in large concentrations could well be a hindrance.

Ice particles could be formed in the atmosphere either by the freezing of existing water droplets or by the sublimation of water vapour direct into the solid state but the evidence points conclusively to the former process as the only one of importance in natural meteorological processes. While there is no evidence that the air is ever significantly supersaturated with respect to water without the formation of cloud, the occurrence of supersaturation with respect to ice without the formation of ice particles is normal. Down to temperatures of $-10°C$ clouds, unless contaminated by falling ice crystals from above, are regularly and perhaps invariably clouds of water droplets alone although at this temperature the air, saturated with respect to water, has a relative humidity of 110 per cent with respect to ice. At lower temperatures clouds are liable to contain ice, while below $-20°C$ pure water clouds are exceptional and the problem is to discover why the droplets freeze. As has already been remarked, pure water drops may be cooled to about $-40°C$ before freezing, and we need therefore to look for some contaminant. The only agent acceptable to the physicist is a nucleus of some insoluble solid which, by virtue of its surface structure and physical properties, may bind adhering water molecules sufficiently firmly and in a suitable arrangement to initiate the ice-crystal development. How such solid particles come to find themselves within a liquid water droplet is not obvious for solid insoluble particles do not act as efficient *condensation* nuclei, but a ready explanation is probably to be found in the occurrence of many 'mixed nuclei'; that is, nuclei of complex particles partly soluble and partly insoluble, which may therefore act first as condensation nuclei and later as freezing nuclei. The chemical nature of the freezing nuclei is variable, but laboratory work has shown that many substances may serve at temperatures below $-10°C$ with soils, sands, and clays, silicates in general, among the more effective; the analysis of nuclei recovered from snow crystals and cirrus clouds has verified the presence of such substances. There is evidence too that organic substances including micro-organisms may be involved on occasion. Laboratory work with a wide variety of chemical substances has not only demonstrated the effectiveness of many nuclei at temperatures below $-10°C$, but has also discovered very few which would initiate freezing at any higher temperature, a result which accounts satisfactorily for the meteorological occurrence of clouds consisting only of water at these temperatures. The iodides of silver and lead seem to be the most active agents so far discovered, becoming effective at temperatures around $-5°C$, and while these are not naturally present in the atmosphere and have no bearing on natural meteorology they do open up possibilities of interfering with nature as we shall see in the later chapter on the artificial control of weather.

Chapter 6

Vertical Convection: cumuliform clouds, showers, and thunderstorms

WITHIN the endless variety of weather experienced in middle latitudes, with – in the ordinary course of events – each day different from its neighbour, the day which dawns bright and sunny is not uncommon at any season of the year but, unfortunately, this is no guarantee of a fine day to follow. People will be heard to say, 'it is too bright to last', or, 'it is bright too early', and likely enough small clouds will gather quickly and by midday large towering cumulus, soon to be followed by showers, will have justified the pessimism. But it is not always so – continuous fine days do occur from time to time – and the problem is to discover the factors which distinguish the one case from the other.

The cumulus and cumulonimbus clouds are the visible evidence of currents or portions of air of no great horizontal extent – a few kilometres at most – rising more or less vertically, for which process the meteorologist has adopted the word 'convection', a technical specialization of a term which, more usually in physics, may mean any mode of fluid movement conveying heat or other property, not essentially upwards or downwards. It would be clearer if we were always to say 'vertical convection' when the stress is on motion in that direction but we shall not attempt to follow such a counsel of perfection.

The essence of the matter lies in the principle of buoyancy associated with the name of Archimedes who, living in Syracuse some 2100 years ago, is reputed to have discovered the law – very appropriately – when in his bath. In modern terms, we may say that the downward force of gravity transmitted through fluid pressure exerts an upward thrust on a body immersed in the fluid. The upward thrust is precisely equal to the weight of the fluid which the body displaces so that there is no net upward thrust, no net buoyancy, on a body of density the same as that of the fluid. A body more dense than the fluid, as a stone in water, will sink, while one less dense, as wood in water or a hydrogen-filled balloon in air, will tend to rise 'by buoyancy'. The upward convection current of meteorology is the buoyant ascent of a portion of air lighter than its environment because it is warmer, and the first step towards understanding convection and convection clouds is to understand how a portion of air may become warmer than its environment.

Perhaps the most obvious mechanism is the heating of the air near the warm ground on a sunny day, for then a portion of heated air breaking away from the ground will find itself warmer than the cooler air a little way above, and will rise a further distance by buoyancy. If we pursue this rather obvious process more meticulously we immediately run into questions of considerable subtlety relating to the size of the heated portions which will break away from the surface, how warm they need to become before they do so, and so forth; but, we shall set these important details aside and pass on to consider the further behaviour of a heated portion as it ascends and changes temperature by adiabatic expansion. The rate of cooling by this process has been mentioned earlier, it is one of the most important quantities in meteorology, 1°C for each 100 metres or 5·4°F for each 1000 feet of ascent, and we need to inquire how far the warmer portion would need to ascend before losing its buoyancy by cooling to the temperature of its environment: obviously everything depends on the temperature of the environment. If the environmental temperature falls off with height by more than 1°C in 100 metres, the rising parcel will never cool to the environment temperature but will continue to gain buoyancy and ascend indefinitely, but if the environmental temperature falls off at a smaller rate, buoyancy will in time be lost. The adiabatic lapse-rate is therefore a critical value.

Cumulus cloud, visible evidence of convection, is first formed if and when the ascending portion of air attains that height which suffices to reduce its temperature to the saturation point, a quite definite height dependent on the humidity of the air in the sample and readily obtained from standard tables if the humidity is known. In conditions when the air mass covering an expanse of country is sensibly uniform, convection currents rising from near the surface in different places may be expected to reach saturation at much the same level – the condensation level – to produce a population of detached cumulus clouds based at that level, a familiar sky over both land and sea, wherever convective clouds occur. A good example is shown in Plate 8. Once cloud has formed within the rising air, further ascent is accompanied by further cooling, but now at the saturated adiabatic rate, less than the full adiabatic value of $1°C$ per 100 metres by virtue of the latent heat made available by continued condensation. In order that ascending cloudy air may continue to maintain its buoyancy, it is necessary only that the temperature lapse-rate of the environment should exceed this reduced value. It is a fact of observation that the lapse-rate in the atmosphere does exceed the saturated adiabatic lapse-rate very frequently everywhere in the troposphere, so that deep vertical convective clouds are commonplace.

In order to discuss the nature of convection on any particular occasion we must therefore first know the temperature of the air mass at all relevant heights, and since convection may on occasion occur anywhere in the troposphere up to heights of 30,000 feet or more, 10 or more kilometres, upper air observations made by sounding balloons are indispensable. In passing, it may also be remarked that as air-mass temperatures change rapidly from day to day and from place to place, it is quite essential to have numerous and frequent balloon soundings if convective processes and possibilities are to be accurately assessed in the current weather-forecast. Given the necessary observational information on temperatures and humidities, it becomes possible to judge to what height convection starting at the surface, or at any other height, would penetrate, with or without convective cloud formation. The procedure is, to imagine a detached portion of air starting an upward excursion from some level and to calculate the temperature it would have during its journey allowing for cooling at the adiabatic rate until condensation begins, and at the smaller saturated adiabatic rate thereafter. If the hypothetical parcel would remain warmer than the environment up to any level, as shown by the available balloon soundings, it may be presumed that convection will in fact develop spontaneously up to that level whereas, of course, if the hypothetical parcel would become cooler than the environment, it would have negative buoyancy and convection would not in reality occur.

Although the calculations mentioned above could well be carried out arithmetically by the aid of tables, they are of a kind which are eminently suited to graphical treatment and, as might be expected, the professional meteorologist has invented highly sophisticated diagrams which only a trained physicist can fully understand. A very useful beginning may, however, be made by quite elementary methods well within the grasp of anyone acquainted with graphical processes of solving arithmetic problems. The basic diagram we may use is on ordinary rectangular graph paper, with height represented in one direction and temperature in the other, as in Fig. 4. The observations obtained from a balloon sounding, giving the temperature at all heights in the air mass, may be directly plotted on the diagram as points which, joined by a line, may be called the environment graph. If the temperature decreases with height, as is usual, the graph slopes upwards from right to left; if the temperature is constant with height (isothermal layer) the graph runs parallel with the ordinate axis; if the temperature increases with height (an inversion) the slope is in the other direction. Nature shows unlimited variety and each case needs special study. For the analysis of convection one must follow the behaviour of air samples moving vertically and changing temperature at the adiabatic rates; to facilitate this, the diagram is provided with two sets of background guide lines. The one set consists of straight parallel sloping lines representing a decrease of temperature with height of $1°C$ per 100 metres ($5.4°F$ per 1000 feet), while the other set consists of curved lines representing a decrease of temperature with height at the smaller rate appropriate to cloudy air. The sloping lines of the first set are straight and parallel because the adiabatic lapse-rate always has the same value, whereas the lines of the second set are curved because the adiabatic lapse-rate for cloudy or saturated air, varies and increases with height.

It is now a very simple matter to construct a graph which will represent the change of temperature with height of any hypothetical ascending parcel of air. Beginning at the point corresponding

1. *Cirrus uncinus*. Dunstable, Bedfordshire. A warm front was approaching and rain arrived wihtin 6 hours.

2. *Cirrocumulus.* Dunstable. Some *cumulus fractus* below.

3. *Cirrus contrails.* Bracknell, Berkshire. Height about 30,000 ft. A warm front was approaching, but the cirrus shown here formed by development from the condensation trails of aircraft.

4. *Altocumulus*. Dunstable at dawn. Height about 10,000 ft.

5. *Altocumulus castellanus*. Malta.

6. *Altocumulus lenticularis*. Taormina, Sicily, Mount Etna bottom left. Height about 10,000 ft. Looking south-westwards. Wind was north-westerly, and these wave clouds formed by the mountains of north-west Sicily persisted for 48 hours.

7. *Altostratus* with *stratus fractus* below. Near Dunstable, Bedfordshire. Height about 8000 ft. Rain was falling.

8. *Cumulus* in lanes. Base about 3000 ft. Near Bracknell.

9. *Stratus*. Ben Bulben, County Sligo, Eire. Height about 350 ft. above ground; 1100 ft. above sea level.

10. *Cumulonimbus*. Luqa, Malta, looking east.

11. *Stratus fractus* with *nimbostratus* above. Bracknell, Berkshire, looking
south-east.

12. Radar scan of advancing showers. The bright echoes show where rain is falling. The faint rings are range markers at intervals of 50 km.

13. Radar equipment on the roof of the Meteorological Station at Singapore.

14. The British Meteorological Office Mark 2B radio-sonde.

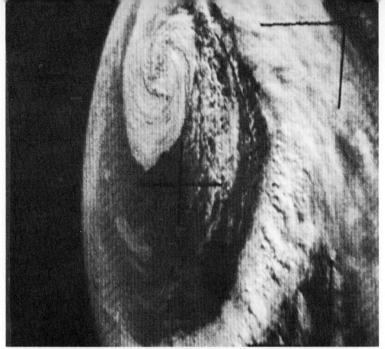

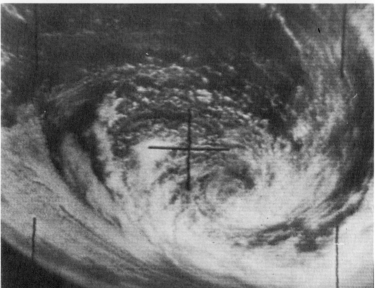

15. The lower photograph taken from US Satellite TIROS VIII. Time
0145 GMT 31 March 1964. The centre of the picture is at about 25° S
111° E, that is off the west coast of Australia. The configuration of the
clouds clearly suggests a clockwise whirl converging in the centre, and it
is in fact the cloud system of a depression which rotates in that direction
in the southern hemisphere.

The upper photograph, centred over the Atlantic near 42° N 60° W, shows
the cloud system of a depression in the northern hemisphere, rotating in
the opposite direction.

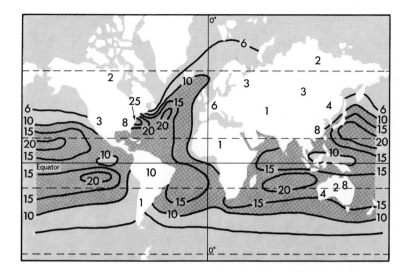

16. Average Yearly Evaporation in decimetres. Oceanic areas with evaporation more than 100 cm per year are shaded. Over the land the patterns are complicated by topography and only a few representative values are shown. Note particularly that evaporation tends to be highest in the tropical zones of the two hemispheres separated by smaller values near the equator itself. Also note the locality of greatest evaporation in the world is off the U S Atlantic coast where the energy is supplied by the Gulf Stream.

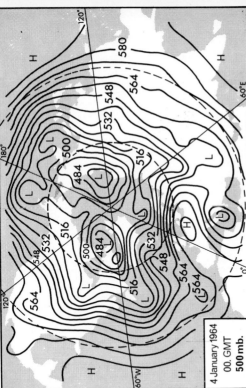

4 January 1964
00. GMT
500 mb.

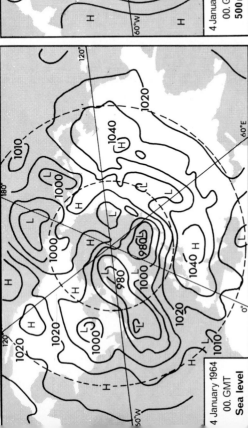

4 January 1964
00. GMT
Sea level

17. *Notes on Charts for Midnight, 3–4 January 1964.*

The 500 mb contours are much more complicated than the average contours for the whole month shown in Plate 18. The generally westerly circumpolar circulation has two quite distinct centres, over Baffin Land and over the Siberian Sea and there are a number of subsidiary lows in lower latitudes associated, but not in any very simple way, with some of the depressions shown on the sea level chart.

The European region is particularly interesting, showing a large surface anticyclone with low pressure both to the north, in the Barents Sea, and to the south in the eastern Mediterranean. There are corresponding features on the 500 mb chart and the upper winds meander in a reversed S-shape – westerly across Scandinavia, northerly across east Europe, easterly to the South, then northerly again into North Africa to join the westerlies of the sub-tropics. This arrangement is often called a 'blocking pattern' or simply a 'block', and the anticyclone over Europe is a 'blocking high', so called because the normal westerly current is blocked and reversed.

Note that on the surface chart some 25 centres are marked, and there are many more small centres which cannot be shown on this small-scale map. This is quite a typical number of centres in the hemisphere at any one time.

with the initial height and temperature, the graph proceeds parallel with the straight 'adiabatics' until the saturation temperature is reached, a temperature which of course is dependent on the original humidity; thereafter, it continues to greater heights by following a curve parallel with the curved 'saturation adiabatics'. If the parcel graph lies to the right of the environment graph the parcel has positive buoyancy (higher temperature) and convection would be expected to occur in nature, but once the parcel graph passes to the left of the environment graph, buoyancy would cease and convection would likewise end. Thus, given the condensation level (which with a little further complication may also be obtained graphically), it is a simple and rapid process to assess whether cumulus clouds would form and how high they would be likely to reach. To compare the graphs with the weather on any particular day can be a fascinating exercise and a few typical examples may be illuminating. In each case we imagine we begin with a fine, cloudless morning and wish to assess the convective activity of the day – assuming no other change to be taking place.

The case when the cloudless sky continues all day may have its morning environment curve as represented by the graph AE of Fig. 4. After cooling during the night, the air near the ground at A is, in this case, colder than the air immediately above. There is, as we say, a surface inversion and evidently no convection is possible at this stage for the graph of air rising from any level would at once pass to the left of the environment graph, indicating negative buoyancy. During the day, however, the surface temperature may rise considerably, say from A to B_1, and convection currents may then ascend as represented by the line B_1P but no further than P, for at this height, 1000 metres, the convection line BP crosses to the cold side of the environmental graph. If the air is too dry to produce clouds at this level (P) the sky will remain quite cloudless all day, assuming of course no other source of higher clouds. Naturally we cannot decide from this evidence alone, whether or not high or medium clouds will drift over the region in the upper winds; but, we can be confident that cumulus will be absent.

With only a slightly increased humidity of the air near the surface it may be that at some point P_1 saturation is reached. The convection graph then follows the line B_1P as represented in

Fig. 4 to the condensation level P_1, and thereafter parallel with the curved guide lines until the environment curve is reached and crossed at Q. Cumulus may then be expected to form at the base level P_1, say 700 metres, and to have tops at 1500 metres, a thin layer quite insufficient to produce showers and likely to mean

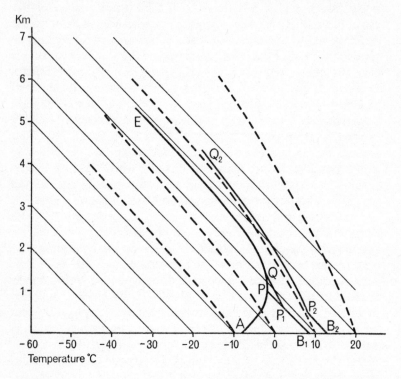

Fig. 4.

a fine day. The case is, however, very interesting because in some circumstances the many elements of cloud between P_1 and Q may join together to form a more or less unbroken sheet covering the whole sky, whereas in other very similar circumstances only patchy clouds will be formed and the day will be generally sunny. To understand the difference it is necessary to consider not only the formation of cumulus elements but also their dissolution for, on a day with broken cumulus clouds, patient observation, continued for 10 or 20 minutes if necessary, will generally confirm what is not otherwise always obvious – that each cloud has but a brief existence and that the apparent uniformity is in reality a continuously changing population of cloudlets. When a cloud has reached its limit of development and upward motion has ceased, there are two distinct processes which may cause it to dissolve in a few minutes. The one process is by mixing with drier air of the environment, and observation suggests that this is normal: the cloud appears to dissolve into thin air at the level where it was formed. There is, however, another process which may cause a cloud to disperse: it may begin to sink and in doing so will warm by compression and dissolve by a process which is the reverse of that responsible for its formation.

The course of a day of convection, when the ascent is severely limited by an inversion of temperature, has therefore very many variants depending more than anything else upon the humidity of the air, not only near the surface, but throughout the convective layer. The convection is an effective mixing agent, and as the day progresses, water vapour – often concentrated near the ground in the morning – becomes more or less uniformly mixed. Thus in the early part of the day isolated clouds will form according to the morning humidity near the ground, but later the behaviour will depend on the humidity as affected by the mixing, and not infrequently this is too low to permit further clouds to form. In these circumstances, a cloudless afternoon may follow a morning of broken convective clouds: although convection continues the air has become too dry. On the other hand, the mixed air may be sufficient to produce saturation everywhere in the upper part of the convective layer, and a morning of convective broken clouds may degenerate into a dull, overcast afternoon. The different cases are not always easy to distinguish in advance, even for the professional weather-forecaster. This is especially so when upper air observations are few, or the humidity is changing by one of many processes – by air drifting in from the sea, by evaporation from the surface, by subsidence, and so forth – and reliable forecasts of the precise cloudiness are then unattainable.

When the lapse-rate of the air mass is greater than the saturated adiabatic rate through a deep layer above the condensation level, convective clouds may be correspondingly deep. Vigorous shower clouds will then develop, perhaps with thunderstorms and hail.

Cases occur when the upper limit of convection is not reached at any height below the base of the stratosphere, frequently above 10 kilometres, say 30,000 feet, in middle latitudes, 15 kilometres in the Mediterranean and similar subtropical latitudes, and even higher in equatorial convection. Magnificent cumulonimbus clouds with characteristic anvil-shaped tops may, in favourable circumstances, be visible on the horizon at great distances, 50 miles or more in England and over 100 miles in lower latitudes.

The variety of possibilities is large with infinite gradations between slight and pronounced instability, shallow and deep convection, but a very important case arises where the lower atmosphere is such as to produce convection of limited vertical extent but is surmounted by air of lapse-rate greater than the critical value. In the state represented in Fig. 4, convection associated with surface heating to the temperature represented by B_1 could produce only flat cloud above P_1, but heating to B_2 with sufficient humidity could produce very deep convection, for the convection graph $B_2 P_2 Q_2$ is on the right of the environmental graph up to great heights. Looking at the matter once more from the realistic point of view of the weather-forecaster, it will be evident that some very precise information, on the upper air conditions and on the amount of surface heating, may be needed in borderline circumstances to recognize the occasion when convection will be sufficiently vigorous to break through the stable layer – which acts as a kind of lid – and penetrate into the upper regions where convection will continue spontaneously. One may perhaps sympathize with the forecaster who can do no more than predict a 'risk of showers' on these occasions.

In the above examples, attention has been directed to situations in which the air mass remains unchanged apart from heating (and evaporation of moisture) at the earth's surface, a very important kind of behaviour which accounts for much of the variety of weather during the course of a day. The cases are, however, somewhat idealized as the air mass itself – and its environmental graph on the aerological diagram – are continuously changing, if only slowly, by all the thermal and dynamical processes which are active in the free atmosphere. To the forecaster looking more that a few hours ahead, the problem of predicting the change in air mass character is often much more critical than that of predicting convection in a known environment, and some very interesting and important developments are possible.

Very frequently there is adequate humidity to form layers of cloud by various processes, but the lapse-rate of the environment is less than the critical saturated adiabatic value and the clouds remain flat (layer clouds) with no liability to break out into upward convective towers. Very frequently also, the critical lapse-rate is exceeded through a deep layer but the air is too dry to produce cloud, and again there is no chance of convective systems. But with the winds blowing, as they normally do, with different speeds and perhaps also different directions at the different heights, both the humidity and the temperatures change slowly or rapidly by 'advection', while the characteristics may also change by slow ascent of the whole air mass over a large area (as in a depression), and also slowly by radiation. It is, therefore, to be expected that in some circumstances a stable arrangement without convection will change, either by increasing moisture or increasing lapse-rate, into an unstable arrangement permitting convection to break out suddenly. Some of the most violent and prolonged displays of convection with heavy thunderstorms, excessive rain, and damaging hail may arise in this way. In a typical case, clouds may gather in the lower atmosphere gradually over many hours through a deep layer with tops perhaps to 10,000 feet, but with little vertical motion and little or no rain, and prevented from breaking through into the higher levels by a shallow layer of small lapse-rate or even an inversion. Cloud layers of this kind actually have an inherent tendency to seal themselves off by radiation cooling from their upper surfaces. But the situation can be in a delicate state of balance, tottering – so to speak – on the brink of catastrophe and liable to violent change if the shallow lid is broken through by any mechanism; perhaps a little cooling of the upper air, perhaps forced ascent at a travelling front. The situation may be such, that all the air in the lower levels, when allowance is made for latent heat, is potentially warmer and lighter than the upper layers and once the process is started there is nothing to prevent the development of large convective cells effectively over-turning the atmosphere, transferring the moist air to the upper troposphere with the deposition of almost all its moisture as rain. An outbreak of thunderstorms of this kind has no particular preference for any one time of day and may well occur during the hours of darkness with spectacular lightning displays.

The structure of convective cloud systems The basic reason why convective clouds develop in some situations and not in others, sometimes shallow and sometimes through a great depth of the atmosphere, has not been too difficult to elucidate in terms of lapse-rates and humidities, the latent heat of water vapour and the principle of buoyancy; but to derive from basic laws all the characteristics of the systems which do develop, raises difficulties of a totally different order. To give a physical description of the confused turbulent motion which characterizes convection caused by a heated ground or relatively warm sea surface is not at all easy but it seems to be something on the following lines.

The convective cells or eddies – samples, parcels, patches, or bubbles of air – which move as entities for a time and then dissipate by mixing in the environment, are both the result of the heating and the mechanism by which the heating is transmitted upwards. Parcels of a wide range of size occur with the dominant size increasing with the height attained, in diameter about half this height, so that a few metres above the ground the dominant eddy is of metre size, whereas at 500 metres parcels up to 200 metres diameter will be found as coherent convective units. In the case when cloud forms these large parcels of well-mixed air reach the condensation level so that distinct and detached cumulus are formed. Energized now by the liberation of latent heat, the cell grows upwards rapidly until all the air of the original parcel has been drawn in from the bottom, whereupon drier air from the environment is entrained and the cumulus begins to collapse and dissolve, the whole process occupying but a few minutes. The process of course leads to an increase in the humidity of the upper air reached by the convection, so that successive parcels are less readily dissipated by entrainment and mixing. On those occasions when the air mass is sufficiently unstable to produce deep shower clouds, the patient observer may distinguish the new bubble which, after entering a mass of inactive cloud debris, towers upwards into a typical cauliflower head, persisting for a minute or two, then collapsing to be succeeded by another, larger in size and more vigorous still. The whole process may continue for up to a few hours, without any rain until sufficient depth is reached for the precipitation process to be initiated, whereupon a shower very quickly develops and rain may be seen to fall out from the base of the cloud in a streaky grey curtain.

Not enough is known about the dynamics of the systems to allow the different varieties of shower to be logically classified, but differences there certainly are. In many cases as for example the type illustrated schematically in Fig. 5, the rain stage marks the beginning of the end for the falling rain rapidly cools the air

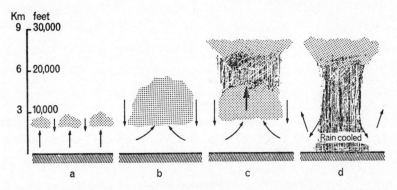

Fig. 5. Stages in the life of a simple shower. After stage (*d*) the whole system dissipates and disappears within half-an-hour.

below the cloud by evaporation and down-currents develop to spread out near the ground and produce the chilly squalls which are so characteristic of showery weather. But at other times, when the shower systems are large, a new and much more persistent structure may be evolved. The whole cloud-mass drifts of course with the general wind, generally the wind at a height of about 10,000 feet, and a kind of front develops between the advancing rain-chilled air moving outwards from the storm and the warm surface air ahead. This new warm air then rises above the cold air and is swept up into the cloud-mass to maintain the whole system as a going concern. The cross-section of the moving storm is then as indicated in Fig. 6. It is no longer appropriate to think of limited parcels of heated air rising by buoyancy through an environment; the better model, is that of warm surface-air being scooped up by the advancing squall-front, reaching the condensation level, and then rising by its buoyancy in a persistent current. A system of this kind may continue in being for an hour or two and bring heavy rain everywhere along its path. It is perhaps not surprising that the shower-front once formed may

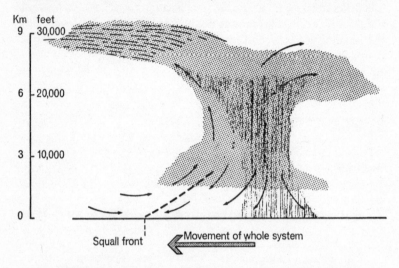

Fig. 6. Persistent travelling shower system. Warm air is scooped up by the advancing squall of rain-cooled air maintaining the system.

develop laterally into something which is almost indistinguishable from the true cold front – described in Chapter 9 – although historically there may have been no difference between the air masses on the two sides and the whole phenomenon is the result of vertical convection.

The electricity of the thunderstorm Wholly mysterious as they were until modern times, flashes of lightning and peals of thunder are in essentials of the same nature as the electric sparks and the accompanying crackles of many kinds familiar to everyone in this age of electricity: sparking plugs of the motor-car engine, electric light, power switches and electric trains, are among the many things which provide examples either by design or incidentally;

but the flash of lightning, kilometres in length, requires not the few hundred volts of an electric power supply, but tens or hundreds of millions of volts. The current is also large, perhaps 100,000 amperes, but it lasts for a very small fraction of a second, and it is for this reason that the whole discharge may be taken without damage by a lightning conductor of a square centimetre in section.

It would be natural to expect that an aircraft in flight through a thunderstorm would be particularly vulnerable to the stroke of lightning, but experience shows that the risk is not very serious, much less so than that caused by the excessive bumpiness, ice formation, and hailstones which all contribute to the risks and make the thunderstorm something to avoid if possible. The metallic hull of the aircraft shields the occupants from the electric discharge which may strike and pass through an aircraft without serious damage, although there are exceptions.

The danger from lightning is, rather surprisingly, more serious on the ground than in the air, for the earth and objects on it are not generally good conductors of electricity and are not in good contact with each other. Consequently, when a stroke of lightning reaches the ground (the leader stroke generally passes downwards from cloud to ground) it shows preference for elevated objects, trees or buildings, and in passing through them to 'earth' frequently has no well-conducting path available and generates heat to an explosive degree. Lofty buildings and isolated trees are particularly liable to be struck, and in regions prone to thunderstorms it is to be expected that any structure such as a church spire will, sooner or later, receive a flash which may easily be destructive. The damage to important buildings was formerly so frequent and serious that the invention, in the middle of the eighteenth century, of the lightning conductor, by Benjamin Franklin, must be regarded as one of the most important among the early benefits bestowed on mankind by science. At the same time, it was the outcome of scientific thinking and experiment which directly proved, for the first time, that thunder clouds were charged with electricity and that the lightning flash was an electric discharge.

The protection afforded by a good installation of lightning conductors is so complete that the risk in this respect need no longer be a serious one, but lightning continues to set problems for the modern world. Examples arise in the control of forest fires, lightning being a common cause, and in the design and operation of electric power transmission lines, lightning being a troublesome source of failures of various kinds. The number of people killed by lightning each year in the United States is about five hundred and, since these mostly occur in the open country to individuals caught out of doors or sheltering in small buildings or under trees, there seems to be little hope of avoiding most of the fatalities which make up, after all, only a minor hazard, insignificant when compared with the dangers on the roads or the innumerable sources of domestic accidents.

Taken by and large, the electrical display of the thunderstorm is more spectacular than important to the ordinary citizen, more a minor nuisance than a factor in a dangerous world. Furthermore, to the best of our knowledge, it has little importance in the economy of the cloud; it is a side-effect, a para-phenomenon, accompanying heavy convective showers but not affecting in any significant way the mechanics of the system, the vertical motions, the cloud, the rain, or the hail. One may say, paradoxically, that about the least important factor in a thunderstorm is the thunder and lightning.

To find a satisfactory explanation of the lightning, and of the separation of electric charge which precedes it has not been easy, not because a source of electricity could not be found but rather because there is an embarrassment of riches, numerous processes but none altogether convincing. The separation of charge leads one to look immediately for carriers moving in opposite directions and nothing could be more ready to hand than the rain moving downwards and the air, with the smaller droplets, moving upwards. The separation must be of the correct sign to explain the distribution of charge as it occurs in the cloud, namely positive charge towards the top, and negative charge towards the bottom, although even this picture is confused by the limited areas of positive electricity at the bottom of the cloud and positive charge on much of the convective rain reaching the ground. The chief problem would, however, be solved if it could be shown that precipitation falling through the cloud mainly acquires a negative charge while allowing positively charged particles or positive ions to be carried aloft in the convective current, provided of course, that the process would separate sufficient charge to explain the amount of electricity involved. There are many possibilities, and when freezing is introduced so that ice as well as water is available, the possibilities increase. It has indeed been shown in the laboratory that electric charge is separated when water drops break up by air friction, when ice crystals collide, when supercooled water droplets freeze and explode, when ice crystals fall through a current of air at a different temperature, even when water drops simply fall through ionized air in an existing electrostatic field. In a situation of this kind it is not to be expected that experts will quickly agree that any one process plays the major rôle in natural thunderstorms and even the view now almost universally held, that ice as well as water is necessary for effective charge separation, is belied by the evidence from airmen who claim to have seen lightning in 'warm clouds' of the tropics, that is in clouds which do not rise above the freezing-level. The position is indeed reminiscent of the theory of rain formation itself in which for a time the freezing of cloud droplets was thought to be an essential stage, although this extreme view is now abandoned. The really large convective system does invariably reach upwards to regions where the temperature is below the freezing-point and, assuming that exceptionally vigorous vertical currents are necessary for the separation of sufficient charge, it is hardly likely that many examples of lightning will be found without freezing also being present: but this in itself proves nothing about the cause of the electric charge. Nevertheless, it is widely believed that the ice phase is a necessity, and the most recent theory put forward by Professor B. J. Mason, depending on the freezing process and supported by both laboratory experiments and detailed theoretical calculations, has gained many adherents. The fundamentals of this theory take one into microphysical processes beyond the scope of this book, but the one essential feature of the argument is that supercooled water drops, when they begin to freeze, become positively charged on the outside surface, negatively charged within, and the surface charge is carried away by the air or by splintering. The theory has been convincingly argued. A remarkable consequence of the charge separation going on in thunderstorms all over the world, a thousand or more at any one time, is that the stratosphere is continuously fed with positive electricity which is distributed over the globe by conduction through the ionosphere, while the earth itself, a free conductor, carries the negative charge. The thunderstorms may in this way be looked upon as continuously charging a leaky condenser, for the balance is maintained by a slow continuous current in the order of a millionth of an ampere for each square kilometre. This minute steady current of fine weather occurs because the atmosphere everywhere is slightly ionized, mainly as a result of bombardment by cosmic radiation, which takes us into aspects of atmospheric physics rather far removed from weather or climate. One should, however, mention that although the fine-weather leakage current is so small, the electric potential necessary to drive it through the very slightly ionized air is very large, hundreds of volts per metre, a million volts through the troposphere, all of which in the ordinary way goes completely unnoticed.

Chapter 7

Radiation and Energy Exchanges

THE interior of the earth is certainly hot, very hot, and its core molten, but for all the effect this has upon the weather it can be completely ignored. Here and there, it is true, in volcanoes and hot springs the internal heat is locally noticeable but, with these exceptions, the heat of the earth, generated by the radioactivity of its mineral substances, is sealed within by the crust and mantle of the earth and leaks out into the oceans and atmosphere at an entirely negligible rate: it is the sun and only the sun that keeps us warm. Were the sun entirely to fail us, an event which the astronomers and solar physicists confidently proscribe for a few thousand million years, the oceans would freeze over in a matter of weeks and the temperature of the earth's surface would gradually fall away to near absolute zero by radiating its heat into the emptiness of space. In fact the earth is continuously losing its heat in this way but the supply is maintained by the intercepting of a little sunshine. Only about one part in 2000 millions of the energy thrown off by the sun, at a surface temperature of about 6000°C and at a distance of 93 million miles, is caught by the earth but this suffices to maintain a tolerably steady level of temperature, getting on the whole neither hotter nor colder, but fluctuating locally and temporarily, from day to night, from summer to winter, and from year to year, according as the gain or loss of heat attains for a time the upper hand.

If the earth were a simple radiator poised in space, the 'black body' of the physicist, it would be possible to calculate a mean temperature of its surface required to radiate heat exactly to balance the intercepted sunshine. The result of the simple calculation (satisfying Stefan's law on the radiant heat absorbed and emitted by perfect radiators) is about 20°C, not far from the actual mean temperature of the earth's surface, but the closeness of the agreement is quite accidental for the earth as a whole is only about 65 per cent efficient as an absorber of solar energy while its effectiveness as a radiator of its own heat is much affected by its covering of atmosphere and blankets of clouds. The temperature of the earth cannot be explained by thumb-nail calculations based on the heat and distance of the sun, but leads to very lengthy and still imprecise calculations which give only a fair approximation to the actual temperatures which we know by direct measurement to be possessed by the earth and atmosphere.

It would be fair to ask at this stage why anyone should be interested to carry out long and tedious calculations to discover what the temperature of the atmosphere should be according to the theoretical laws of radiation, when it is much more accurate to take a thermometer and make measurements, but the point of course is that the truth of the calculated results is the verification of the theory. If we are to understand the problems of long-range forecasting or climatic change, or to judge what would happen to the climate if the conditions were altered naturally or artificially, there is no substitute for a well-proven theory.

Considering first the sunshine, or solar radiation, the energy is spread over a wide band of wave-lengths and, in this environment, biological evolution on earth has led to organs, the eyes, remarkably sensitive to the radiation and in some cases highly discriminatory as to wave-length. Animals have, in other words, developed the power of vision, the sense of sight, and colour discrimination is the subjective recognition of varying wave-length or mixture of wave-lengths. When we say that the middle band of the solar spectrum is the 'visible' band, and the 'visible range' is the colour spectrum from the red to the violet, we are really describing the sensitivity of the human eye and introducing extraneous biological considerations into the physical problem. Similarly, when the invisible band with a longer wave-length than the red is called infrared and that with shorter wave-length than the violet is called ultraviolet, the distinctions are subjective and arbitrary having no particular significance in physics. In so far, that is to say, as the weather and climate of the earth would be the same as it is at present and would rest on the same theory or natural laws even if there were no people to enjoy the sight, it may be said that vision and light and colour can be ignored, and discrimination by wave-length and energy is sufficient. Since the great Newton himself was at some pains to make clear this distinction between physical fact and subjective response, it is not a matter of surprise that confusion lingers on among lesser mortals, and that it is not obvious to everyone that the sun's light is radiant heat differing only in wave-length and energy from the radiant heat given out by any object whatever, according to its temperature and constitution, whether visibly shining by its own light or not.

Practically the whole energy of the solar spectrum lies in the band of wave-length from 0.2μ to 4.0μ ($1000\mu = 1$ millimetre) of which rather less than half, between 0.4μ and 0.7μ, is visible to the human eye. Until very recent times when, by the use of rockets and earth satellites, it has at last become possible to examine the sun's radiation as it arrives from space, observations of the total intensity and spectral distribution have been more or less affected by the atmosphere through which the radiation has passed on its way to the measuring instruments. Yet by piecing the evidence together from many observations in various circumstances, especially from mountain observatories, it has been possible to arrive at what we believe is a pretty good idea of the truth. Even so, experts in 1963 were debating the value of the solar constant – the mean intensity of the sunshine – within a range of 4 per cent, and much of the detail remains to be studied. Nevertheless, from the point of view of climatic and weather studies, we are more certain of the quantitative nature of the solar radiation out beyond the atmosphere than we are of perhaps any other quantity which we need to consider once the radiation has entered the atmosphere and become involved in the complicated processes which then ensue. Few of the components are accurately measurable and the measurements that have been made have been almost confined to a relatively few well-equipped observatories, leaving the wide areas of oceans and undeveloped lands virtually untouched: radiation climatology has been a neglected subject.

To give an impression of the nature of the problem, it is worth while before putting in any numbers to run over the factors and processes which are involved. The sunshine falls not upon a black body but upon an atmosphere of mixed gases and clouds and allowance must first be made for reflection back to space of a large proportion of the energy which then takes no part in terrestrial affairs. The most effective reflectors are the clouds which stand out very clearly as white shining patterns in photographs taken from above such as those illustrated in Plate 15. By comparison, the non-cloudy air is highly transparent but not entirely so and a certain amount of light is scattered by the molecules and some returns to space: it is this process of scattering which gives brightness to the day-time cloudless sky as seen from the ground, and the blue colour is the result of selectivity in the scattering. The light which penetrates to the ground is also partly reflected, and although in the absence of snow or ice the proportion is not large it must be included in the account. The fraction of the light returned to space in this way from cloud, air or earth's surface is known as the albedo, on average say 34 per cent.

The light not reflected is absorbed and converted into sensible heat tending to raise the temperature of the substance concerned, the air, the cloud, the land, or the sea as the case may be. These bodies at the same time radiate energy themselves, wholly invisible radiation for their temperatures are nowhere above about 30°C and their characteristic radiation has long wave-lengths which hardly overlap even with the infrared part of the solar spectrum. The picture is then of the land and sea radiating upwards continuously, some of the radiation being quickly absorbed by the water vapour or carbon dioxide in the air and some passing through the atmosphere into space except when intercepted by clouds, very efficient absorbers of all the radiant heat which reaches them from below. At the same time, the air and clouds also radiate in all directions and much of the energy is returned to the ground or sea surface while some eventually escapes to space.

For a number of reasons, these radiation processes become excessively complicated. First, in the atmosphere it is the clouds and the water vapour, and to a less extent the carbon dioxide, which act as absorbers and radiators while the main constituents of the atmosphere, the nitrogen and oxygen, are in this respect inert although they share in the temperature changes by mixing. The heat capacity or specific heat of the air depends mainly on the nitrogen and oxygen but the process of absorption or radiation depends on the clouds and on the relatively small amounts of water vapour and carbon dioxide. Next, these gases are selective absorbers and radiators, that is they absorb some wave-lengths very readily, others only slightly and yet others not at all. Clouds, on the other hand, absorb and radiate with almost 'black body' efficiency in these terrestrial wave-lengths. The surface of land or sea, whatever its colour in visible light, even when white with snow, may also be regarded as 'black' in the sense that it radiates heat appropriate to its temperature completely freely. Further complications arise from the fact that solar heat striking the sea penetrates to considerable depths (water is transparent) and is stored there, whereas over land the storage is slight; moreover energy in the sea is not static but is carried by ocean currents so allowing energy received in one place to be returned elsewhere. But the troubles do not end here, for radiation is not the only process which transfers heat between the surface of the earth and the air above; quite as important is the evaporation of moisture requiring the supply of latent heat which, being supplied by the surface, is carried by the vapour and is released to add heat to the atmosphere only when the vapour is condensed to clouds and precipitation. Considerably less important but still significant is the heat transferred to the air by contact, that is by conduction.

It will readily be conceded from this involved description of events that to calculate from pure theory the changes of energy and changes of temperature all over the earth and throughout the atmosphere as these quantities vary from day to day would be a gargantuan task. So far it has not been attempted in detail, although with the aid of the most advanced electronic computers the problem is not beyond approach and something of the kind must be undertaken if forecasting by calculation is to take account of all the significant factors. J. Smagorinsky, of the United States Weather Bureau, was recently engaged in a brave undertaking on these lines using the resources of a Stretch IBM computer. It is reasonable that a few years of work for a team of scientists are needed for such a computer programme to be devised and fully tested, and the outcome of each enterprise is awaited with much interest by dynamical meteorologists everywhere; for these are projects not to be lightly undertaken, and unlikely to be duplicated elsewhere for some time to come.

81

It has, however, for many years been well within the scope of theoreticians to attempt to draw up balance sheets for the different components of the energy exchange for the world as a whole. We take as a starting point the fact that neither the earth nor the atmosphere is becoming significantly warmer or colder with the passage of the years, with the implication that in the long run the different processes must be in almost perfect balance. It would take too long to attempt to explain all the ingenious methods by which different research workers have attempted to arrive at reasonably convincing estimates of the various quantities, making use of cross-checks where these are possible, but a few points may be mentioned. Evaporation from the land and sea, which is a major term in the loss of energy from the surface, must over a long period be balanced very accurately by the total precipitation, including both rain and snow, for the atmosphere itself is unable to hold much moisture in store, not more than the equivalent of about ten days' rainfall. A cross-check is therefore provided by making for the whole earth independent estimates of evaporation and rainfall, neither of which, unfortunately, is at all easy to determine. Rainfall, at least, one might think easy enough to measure with an ordinary rain-gauge, and there are tens of thousands of rain-gauges regularly in use all over the world, but the sad truth is that no one has yet devised a way of measuring rainfall with any reasonable reliability from a ship or a buoy at sea, and most of the earth's surface is ocean. Rain generally falls in disturbed windy weather when the amount of water collected by any gauge yet devised is grossly affected by the movement of the ship and by the eddying winds around it. In experiments carried out by the British Meteorological Office rain-gauges were exposed in many different positions on board a ship, swung on gimbals on the yard-arms and so forth, but it was not possible to get even roughly equal readings and it was quite impossible to know what was the true rainfall. Salt spray mixed with the rainfall naturally does not make things any easier but this, if it happens, is not the major difficulty that might have been expected, for by measuring salinity the proportion of salt water mixed with the pure water of rain is readily determined. The difficulties are such that at the present time I believe no country in the world is attempting regularly to measure rainfall on board ship and all our estimates for the wide oceans rest on uncertain evidence. Much weight is often given to the rainfall measured on small islands, taking this as valid for the ocean in the vicinity, but anyone who has lived for a time on a small island or indeed anyone who has noticed from the air how clouds gather round oceanic islands will be very suspicious of the assumption. Quite recently, G. B. Tucker of the Meteorological Office used an ingenious method of graduating ships' observations by studying the frequencies with which ships regularly reporting the weather on the Atlantic had observed rainfall of different intensities, slight, moderate or heavy, and comparing them with observations from land stations where the rainfall amount was accurately known. Far from giving one confidence in previous estimates of rainfall over the ocean, Tucker's result suggested much smaller amounts and, in spite of the fact that no one would wish to put great faith in the new method, it has as much chance of being correct as any other yet devised. And, if rainfall, which is at least observable, cannot be measured, evaporation, an invisible process, would seem beyond hope; but theoreticians are nothing if not venturesome and they have devised formulae relating evaporation with the strength of the wind, the sea temperature, and the humidity of the air, which have obviously got a lot to do with the process. The formulae may be tested to some extent by applying them to small expanses of land water, even water in tanks, from which the loss of water by evaporation can be measured but one may be excused for being sceptical about the results especially when, to obtain evaporation from the oceans, one has to base the calculations on winds estimated by the Beaufort Scale and humidities obtained from wet-bulb thermometers exposed on the ships. Ships' observations of weather are vitally important for weather-forecasting, so important that without them forecasting for the British Isles, for example, would often be blind guesswork, but to expect their routine observations of wind and humidity to be sufficiently accurate to be reliable indicators of evaporation from the sea surface is really to go too far. Nevertheless, it is on such evidence that estimates must be made. Sometimes textbooks may give the average annual rainfall all over the earth as 100 centimetres, say 40 inches, and the very roundness of the figure hints at its imprecision. Quite recently a new estimate led to the figure of 84 centimetres and, frankly speaking, no one knows the truth although as it is strictly limited by the energy available for evaporation we may be fairly confident that the value is not widely outside this range.

All the other terms in the energy balance have been estimated by various authorities and Budyko of the USSR has gone so far as to produce an atlas of the world giving the geographical distribution of all his estimates; and the author's reputation for scholarship is sufficient to give one confidence that the evidence from many sources has been most carefully weighed. As an illustration of the kind of balance sheets that have been struck we give in Fig. 7 (p. 85) a diagram after Houghton (1954) as quoted by Byers in his standard textbook on General Meteorology. By reference to the accompanying legend, the diagram is self-explanatory and will repay the effort of a few minutes' careful study item by item. The balance sheets for the earth as a whole in its exchange with space, for the surface of the earth, and for the atmosphere alone may be read off from the diagram to give the quantities in the following table.

Balance to space		Surface balance		Atmospheric balance	
In	*Out*	*In*	*Out*	*In*	*Out*
100	$-$S 9	D 24	W_E 101	W_E 101	$-W_A$ 48
	R 25	N 17	E 23	E 23	$+W_A$ 105
	$-W_A$ 48	$+$S 6	C 10	C 10	
	T 18	$+W_A$ 105	T 18	A_S 19	
100	100	152	152	153	153

For the understanding of weather and climate, the atmospheric balance is perhaps the most significant and the results are very remarkable. We notice first that although the air receives 101 units by radiation from the ground or sea below it throws back 105 units to the surface once more so that in this respect it is hardly true to say that the earth's atmosphere is heated from below: in the long-wave radiation exchange the atmosphere acts,

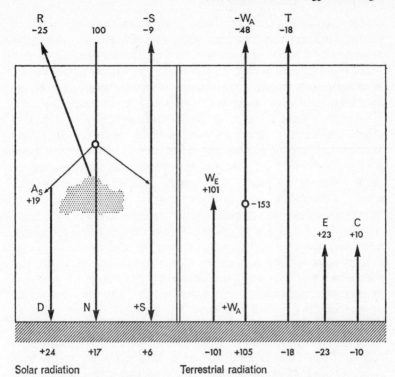

Fig. 7. The heat balance of the earth and its atmosphere as summarized by Houghton.

A_S =   solar radiation absorbed in the atmosphere, 19 units

R   =   reflected solar radiation (depicted as coming from cloud, although some comes from the surface), 25 units

−S   =   solar radiation scattered outward, 9 units

+S   =   solar radiation scattered downward (sky radiation), 6 units

D   =   direct solar radiation reaching the earth, 24 units

N   =   diffuse solar radiation reaching the earth through clouds, 17 units

W_E =   emission from the earth in the absorbing portion of the water-vapour spectrum, 101 units

$+W_A$ =   downward flux of infrared radiation from the atmosphere in the absorbing portion, 105 units reach the ground

$-W_A$ =   upward flux of infrared radiation from the atmosphere in the absorbing portion, 48 units leave at top

T   =   emission from the earth in the transparent region, 18 units

E   =   latent heat carried to the atmosphere in the hydrologic cycle (evaporation, condensation, precipitation), 23 units

C   =   heat transported upward by convection or turbulence, 10 units to the atmosphere it seems, as an excellent blanket to the surface allowing little heat to escape. Ten units are assigned to surface heating of the air by conduction, a small figure which later authorities believe is still over estimated and should be more like 5 units, and the main net source of heat to the air is either by direct absorption of the sunshine, 19 units, mostly at low levels by cloud and water vapour, or by the latent heat liberated from condensed water in rainfall, again of course mainly at low levels for heavy clouds are formed and rainfall is liberated mainly below a height of 10,000 feet. It is, however, a point of extreme importance to a proper understanding of climate that, overall, the atmosphere gains heat mostly in rainfall, a fact which also needs a good deal of explaining when we return to common sense and remember that rainy weather is not at all warm weather. At a later stage when discussing weather phenomena or the general circulation of the atmosphere, this paradox will be further clarified but we may anticipate the discussion here by noting that these overall averages hide a vast amount of geographical and synoptic pattern. In some places, and at various times, there is a large imbalance of energy which is equilibrated by the movement of the air carrying energy from place to place, a matter of dynamics. In particular, when latent heat is liberated in rainfall, the air actually receiving the heat is rising to higher levels so that its temperature does not in fact increase: the energy is used to lift the air and is converted to gravitational potential energy. But elsewhere in the dynamical system air must sink to replace the ascended air and it is this air which actually gets warmer. In other words, when water condenses in the atmosphere and rain falls the heat released is used to drive a convective engine which transfers the energy to warm the air in another place. In a sense then, we may say that the warmth which we experience in the sinking air of fine-weather anticyclones comes in no small part indirectly from the rainfall in depressions and there may be a grain of comfort for those who suffer from a cool wet summer in England in the thought that they are helping to provide good weather for holiday-makers on the Riviera: or is the suffering thereby only aggravated?

Chapter 8

Synoptic Meteorology –
a World-wide Organization

SCIENTIFIC terminology, curiously enough, is often arbitrary or even accidental and nowhere more so than in the names of the different sciences and divisions of sciences. Who, for example, could guess the accepted distinctions between geology, geophysics, and geography; who would imagine that the word meteorology could be used for the study of the atmosphere excluding, incidentally, the study of meteors; who could begin to imagine what synoptic meteorology may connote from a knowledge of the original derivations and meanings of the classical words? There is usually some kind of logic in the early use of a name, sometimes bad logic as in the recent attempt, which may be successful, to fix the name 'aeronomy' to the study of the outer fringes of the atmosphere, but after a time the name becomes little more than a label, a proper name, which remains unchanged however much the subject evolves. The name 'synoptic meteorology' is of this kind and we have also 'synoptic charts', 'synoptic observations', 'synoptic meteorologists', even 'synopticians', the meanings of which are discovered not by reference to a dictionary, unless it is very complete and up to date, but by finding the thing in question and getting acquainted with it. The idea behind the use of the word 'synoptic' is the taking of an overall view of the weather in a geographical sense as distinct from a local view from an observatory, a statistical view from climatic records, or an experimental view from a university laboratory. Thus a map showing weather-conditions at any time is a synoptic chart, a fact which has led many people into the error of identifying the word synoptic with synchronous. Early synoptic charts were in fact often correctly known as synchronous weather-charts but no one, happily, was tempted to call the study of these charts 'synchronous meteorology' or to call himself a 'synchronous meteorologist' or even a 'synchronologist': we should, I think, be grateful for the word synoptic.

Synoptic meteorology is then distinguished by its characteristic method of approach rather than by its scientific content. The primary tools have been weather maps showing the geographical distribution of pressure, temperature, wind, humidity, and weather generally, as inferred from the observations made at the observing stations and at agreed times, generally several times each day. The task has been to study the ever-changing distributions, to describe and to understand them. It so happened that long before it was possible to explain or understand the moving geographical patterns of weather, it was possible to predict the changes by sheer empirical extrapolation and with sufficient reliability to make the predictions interesting and useful. So it was that weather-forecasting was born about a hundred years ago, was introduced into every advanced country, and has been practised ever since, almost exclusively in official national institutions such as the Meteorological Office in Britain or the Weather Bureau in the United States. Synoptic meteorology then became the main preoccupation of official weather services, relatively large funds became available, especially with the advent and expansion of aviation, for building up the system of observations and communications. Relatively large numbers of staff were required to operate the services, to construct the synoptic charts and to maintain a continuous forecasting watch, becoming in time a twenty-four-hour watch through every day of the year. Indeed, meteorology in the guise of weather-forecasting evolved from a rather obscure branch of physical science into something approaching big business and, bearing in mind that the methods of forecasting were for a long time largely empirical and of no deep scientific interest, it is not surprising that the attitude of the universities in most countries was discouraging, almost cold. There was certainly a phase during and between the two World Wars when weather-forecasting services, especially for aviation, grew very rapidly and synoptic meteorology developed out of all recognition but, being backed by no proportionate effort of scientific research and being utterly ignored in most universities, the methods remained for the most part empirical; experience counted for more than scientific eminence, and scientific understanding advanced unnecessarily slowly. It would be wrong to paint a picture of scientific stagnation for there was a good deal of healthy thinking going on within the weather services and it was between the wars that the concept of fronts – warm fronts, cold fronts and occlusions – was introduced by the Bergen school of meteorology, together with the associated ideas and methods of air mass analysis. During a period when, apart from wind observations made with pilot balloons in clear weather and the rare aeroplane ascent, upper air observations were almost completely absent from the current daily weather records, the utmost use had to be made of observations made on the ground or by ships at sea in order to infer the motion of the atmosphere in three dimensions. By the outbreak of World War II in 1939, synoptic meteorology was well advanced and weather-forecasting was ready to contribute substantially to the war effort. This is not the place to remind ourselves of the tragic necessities of war, but those who protest at the prominence given in national meteorological services to synoptic studies and weather-forecasting, with the relative neglect of other parts of meteorology, must concede that the forecaster was a key man in the planning of military operations. The large-scale heavy bombing attacks on Germany, which kept Britain's offensive spirit alive through dark years, could hardly have been carried on without him, and the final landing of Allied troops in Normandy in 1944, the beginning of the end, was an operation planned to be dependent on the advice of weather-forecasters. It was in fact carried out in the light of the predictions made by American and British forecasters in collaboration, and the critical value of the advice provided has been fully recognized by the responsible authorities. Weather-forecasting came out of World War II with honour, and the relatively large resources granted for the further advancement of the science since that time may in no small part be an indirect acknowledgement of the contribution then made.

Synoptic meteorology today can quite conveniently be divided into organization on the one hand, and science on the other, and both aspects must receive attention: which is the more interesting or the more important is a matter of opinion but, if both are not good, synoptic meteorology cannot thrive. From its earliest days the elaboration of synoptic meteorology has been dependent on international agreement and the origins of the International Meteorological Organization can be traced as far back as 1853, being in essentials a conference of directors of weather services supported by various committees of specialists. In the tidying-up operations after World War II, the Organization was dissolved and reconstituted as a Specialized Agency of the United Nations with a minor change in name, World Meteorological Organization for International Meteorological Organization, and with structural changes of some importance, but continuing the traditions and very much the procedures of the old and well-tried body. The main difference was that WMO became 'inter-governmental', making agreements on technical matters with the standing of political treaties, and with much of the solemnity that surrounds such instruments; but in spite of the handicaps of official machinery and the unavoidable intrusion of political bias on occasion, the Organization has been saved by the good sense of professional scientists, or perhaps even more by the inevitable logic of science and technology, to maintain a quite remarkable record in the continuous adaptation of its agreements to meet the changing circumstances of world meteorology. There must be very few amongst the hundreds of meteorologists that from time to time have taken part in the deliberations of WMO who have not had a feeling of pride in its achievements. At its meetings of Commissions, Associations, Panels of Experts, and Working Groups, and of its Executive Committee and its four-yearly Congress, the East and West sit down with the rest of the world to find acceptable solutions to technical problems and rarely break up without making good, sensible progress. In effect, all the international organizational arrangements of synoptic meteorology as outlined in this chapter, and including specifications in great detail to be found in a shelf-full of WMO reference books, have been discussed, hammered into shape and agreed around a WMO table, and there is no sign that the machinery will not work effectively in the future. The WMO has its headquarters in Geneva, a handsome modern building not far outside the gates of the Palais des Nations. Its budget for the four years beginning 1 January 1964, as agreed by its Congress in 1963, stands at over 5 million US dollars, not including very considerable sums deriving from elsewhere within the United Nations and disbursed by WMO in Technical Aid to developing countries, in training of personnel, and in the support of scientific meetings. The permanent secretariat, at present headed and vigorously led by a Welshman, Mr D.A. Davies, provides us with at least one body of international civil servants who can look back at any time on a record of achievement and look forward with confidence to the solution of current problems with no more political acrimony than may occasionally lend a little spice to the proceedings. WMO may be accounted a credit to the world.

By making a précis of the material contained in the WMO reference volumes it would no doubt be possible to prepare a highly informative chapter on international meteorological arrangements but the account, necessarily full of facts and figures, of schemes and schedules, might be dull reading and, as this is not a technical reference work, it may be better to give a general sketch rather than a blue-print. I shall adopt this more congenial alternative.

The goal towards which all the arrangements are aimed is to make available to weather-forecasters, climatologists, and other professional experts at meteorological centres everywhere, all the information about the weather and the state of the atmosphere in general that they will find useful, from all parts of the world in which they are interested, as frequently as is justifiable and with the minimum of time lost in the communication; always, that is, within the scope of the resources which are made available, mainly, it must be remembered, from the national budgets of the co-operating countries. The complementary task of the forecasters and other specialists is of course to make the best use of the information in the preparation of summary reports and forecasts and in the distribution of advice to whomsoever may be concerned: airmen, seamen, or landsmen, engaged in travel or transport, industry or agriculture, sport or other recreation. An outline of the organization needs then to take us through all the stages from the making of the basic observations to the distribution of forecasts and other advice.

The primary contributors are in a very real sense those who

make the observations at the many stations scattered all over the world and the reference works list over 7000 synoptic reporting stations on land and over 4000 selected merchant ships which make observations when at sea. A typical full report will include barometric pressure, air temperature, dew-point temperature as a measure of humidity, wind speed and direction, the forms and amounts of the clouds in the sky with estimates of their heights above the station, the horizontal visibility and a specification of the weather chosen from a hundred possibilities from clear sky to thunderstorms and including drizzle, rain, snow, and other phenomena classified by intensity. An observing station should be regarded then as a standardized scientific observatory, and the precise regulations on the instruments to be used, on their care and maintenance, correct use and exposure, and on the methods of making and recording the observations are the subject of technical handbooks. To illustrate the system the case of air temperature may be chosen. The thermometer, of adequate precision and accuracy, is mounted within a ventilated shelter or screen of specified construction, at a height of 1·5–2 metres above the ground, in a well-exposed site away from buildings; it is read and recorded to the nearest tenth of a degree and the result is rounded off to the nearest degree for synoptic reporting. Supplementary thermometers record the maximum and minimum reached over a period, and autographic instruments are often available for maintaining a continuous graphical record. At sea, special precautions are necessary to ensure that the temperature is that of air fresh from the sea, unaffected by ships' funnels, ventilators, or heated decks. It would be tedious to go further but wrong not to have made the point that routine weather observing is a skilled task demanding technical understanding and specialist training: its importance is often insufficiently noticed.

The observations are made not once each day but many times, even every half-hour at airfields and typically every third hour from midnight universal time (Greenwich Mean Time) onwards. The main hours for observations exchanged internationally are 06, 12, 18, and 24 GMT. When listening casually perhaps to a weather forecast on the radio it may cross one's mind that during the same few minutes before the hour many thousands of men and women all over the world have been engaged in almost the same operations, guided by instruction manuals of almost identical scope printed in every needed language. But international uniformity does not end here. The reports must be collected at sub-centres, national centres and regional centres by radio, telephone, and teleprinter and internationally exchanged through the most elaborate world-wide network of specialized communication networks, radio and land-line, that the world has yet evolved for a specific peaceful purpose. To know what the weather was at the North Pole or in China or in mid-Pacific a little while ago is no substitute for knowing what it will be tomorrow near one's home but it is something to marvel at and, in the event, may be of practical use.

So far in this sketch no mention has been made of upper-air observations and to the many world observers at 7000 or so synoptic stations must be added small teams of upper-air observers at nearly 700 stations. The buoyant hydrogen-filled balloon is the standard vehicle released to rise at perhaps 1000 feet per minute and attaining heights of anything up to 120,000 feet. The active air-borne instrument already briefly described in an earlier chapter is the radio-sonde, essentially an automatic reporting station measuring pressure, temperature, and humidity by means of light-weight sensors and transducers, eventually converting the meteorological measurements into specific radio signals transmitted to the receiving station on the ground. Modern developments in automatic control of rockets, earth satellites and space vehicles have familiarized everyone with electronic sophistication which may seem to put the radio-sonde into the class of children's toys but, as a precision instrument produced for a few pounds and used daily in its thousands for exploring our planetary air-space, it performs a task which has no near parallel throughout the range of human activity. By tracking the balloon, employing radar or radio methods, winds are obtained as well as the instrumental readings so that each balloon ascent gives a vertical, or near vertical, profile of the state and motion of the atmosphere through 90 per cent or even 99 per cent of its mass, and establishes synoptic meteorology as a direct three-dimensional observational science within the broad area of fluid mechanics: we explore the atmosphere full-scale as we might the air currents in a laboratory wind-tunnel.

The upper-air information goes to swell the messages which need to be sent around the world on meteorological communication channels, and mention must be made of the system whereby all language difficulties are overcome by converting the messages into numerical codes of five-figure groups, a natural system when most of the information is essentially quantitative and numerical. A weather message will consist of an indicator group defining the type of message, synoptic, upper air, and so on; a station identifying group which in five figures gives us 100,000 possibilities, many more than the total number of reporting stations of all kinds. Then follows a series of groups giving the scientific information, three figures for pressure, two for temperature, and so on, the nature of the information being defined by position in the message. A central station handles every day something like a million 5-figure groups, all of which are scrutinized. The international, indeed world-wide, agreement on every detail of weather codes is a truly remarkable achievement attained by untold patience and the burning of much midnight oil; and when we recall that every economy is multiplied many millions of times in the handling of weather messages, the time and effort need not be grudged, although the appreciation of the beauty of the final result calls for an aesthetic which is not granted to everyone and which need not detain us here. There is, however, an aspect of these 5-figure codes which is beginning to demand more attention. The codes were designed to be not only economical but adapted to the peculiar predilections of the human mind in recognition, sorting, and coding. As we move steadily towards automation, to machine sorting, mechanical plotting and electronic computing, quite different criteria arise and the ingenuity of the old codes may be irrelevant or, even worse, a source of error and confusion. In particular, human failings such as the omission of complete groups or erroneous coding are often obvious at a glance to an equally fallible but equally intelligent human recipient and the harm done is minimal; but a machine must be programmed specially to cope with each contingency and an undetected error in the make-up of a message may ruin a calculation. The design of weather messages is about to enter a new phase and thinking about it has barely begun.

The fact that highly sophisticated machine methods at some few centres must for a long time to come live alongside the old subjective methods and must feed upon the same raw data, is not calculated to make the transition an easy one. It may well be that weather messages will need to be segregated and even duplicated in machine and human versions, and the World Meteorological Organization has many all-night sittings to look forward to in the reconciling of conflicting claims in codes and schedules.

In what remains of this introductory chapter to synoptic meteorology, a geophysical industry it would seem to be as much as a branch of science, I shall say no more of the objective mechanical methods of the future but shall describe the main features of the techniques of conventional synoptic meteorology. Fortunately, we are away to a good start in the knowledge that no one taking up this book is likely to be unfamiliar with the traditional and still basic device of the synoptic meteorologist, that is the surface synoptic-chart or weather-map of which examples, of limited scope and simplified content, are regularly reproduced in newspapers and on the television screen.

The weather-map is quite simply a geographical map of the weather built up from the evidence provided in the synoptic weather-messages: it is a kind of ordnance survey of the weather. On a geographical survey map special features, churches or railway crossings are denoted by special symbols; lines showing the run of the railways, roads, and rivers have their conventional markings; wooded areas, open country, lakes or seas are distinguished by different hatchings or colour shadings; the height of the land above sea-level is brought out more clearly by the drawing of contour lines for suitable intervals of height: the complete map contains, in these various ways, a great deal of information which is apprehended immediately by the regular map reader. The weather-map is analogous in many ways. Special symbols are employed to denote thunderstorms, hail, types of cloud, and so forth; lines with conventional markings or colours are used for fronts, trajectories, or the tracks of depressions; hatching or colour shading gives prominence to the areas experiencing rain or fog or, alternatively, may be employed to distinguish air masses of tropical or polar origin; and, finally, lines of equal pressure, called isobars, entirely analogous to contour lines, are drawn at convenient intervals to bring out clearly the geographical distribution of pressure as revealed by the readings of barometers. Corresponding with the hills and valleys shown by the contour-map, we have anticyclones (regions of high barometric pressure), depressions (regions of low pressure) as well as many other configurations of isobars which are often described by terms taken straight from the language of topography as, for example, ridges, troughs, and cols. A completed weather-map fully analysed, that is with all the features of interest diagnosed and represented by the conventional markings contains a great deal of information which may be taken in almost at a glance by the expert to provide a mental picture of the atmospheric structure such as an architect may gain of the structure of a building from a set of drawings. The analogy with the architect is, in some respects, not a bad one for in both cases the structure to be represented is in three dimensions requiring plans or maps for different levels to give a full representation. In synoptic meteorology the single 'surface weather-map' was for long the only chart for the sufficient reason that data were too few to permit of upper level analysis, but with present networks of observations of upper winds, temperatures and humidities, a true three-dimensional representation has become standard practice.

On the surface weather-map isopleths are drawn for barometric pressure reduced to mean sea-level but in the upper air a rather different convention has been adopted and we prepare maps for standard pressure levels rather than for standard heights. The pressure at mean sea-level averages a little over 1000 mb so that at a pressure of 500 mb in the upper air about half the atmosphere by weight is above and half below: 500 mb is, in this sense, the middle of the atmosphere and is to be found at a height somewhere near 18,000 feet or 6 kilometres. The height of the 500 mb level shows a geographical distribution varying of course with time and its configuration is shown by drawing what are strictly contour lines at convenient intervals of height, say 200 feet or 60 metres. The same can be done for many pressure levels and the series 700 mb (10,000 feet or 3 kilometres), 500 mb, 300 mb (near 30,000 feet or 9 kilometres), 200 mb and 100 mb (near 54,000 feet or 17 kilometres) are a commonly chosen set. By virtue of this

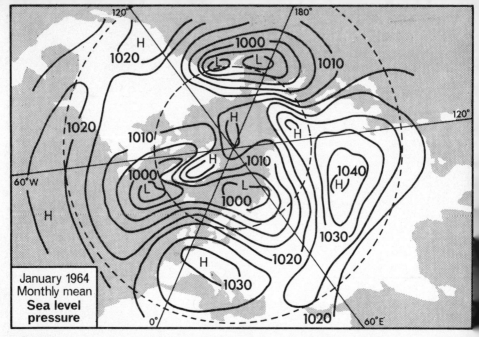

January 1964
Monthly mean
**Sea level
pressure**

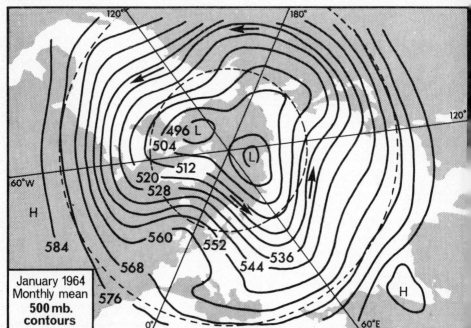

January 1964
Monthly mean
**500 mb.
contours**

18. *Notes on Mean Monthly Charts for January and July 1964, for sea
level and 500, 300 and 100 mb (about 18,000, 30,000 and 50,000 feet)*

This series of charts for conditions averaged through the two months of
January and July 1964 illustrates the important differences between a

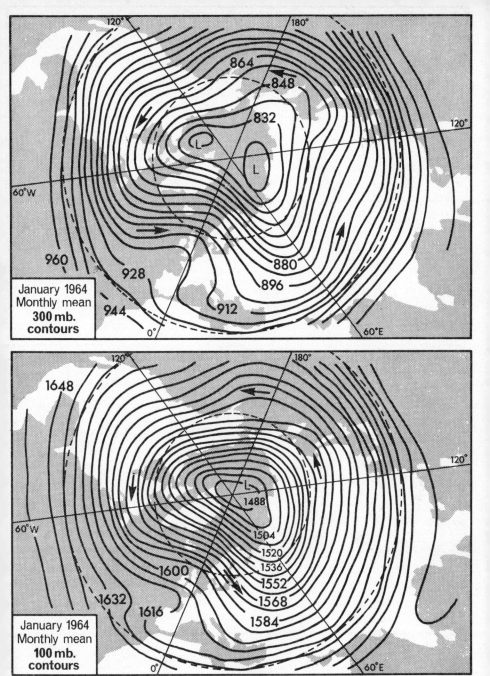

January 1964
Monthly mean
300 mb.
contours

January 1964
Monthly mean
100 mb.
contours

winter and a summer month in the structure of the cirumpolar vortex
of westerly winds. In the winter month the vortex is notably the stronger
(contour lines more closely packed) and considerably the larger. Thus the
lines surrounding the Pole continue down to the tropics in winter, but in
summer not generally south of about 35° N. The contrast is especially

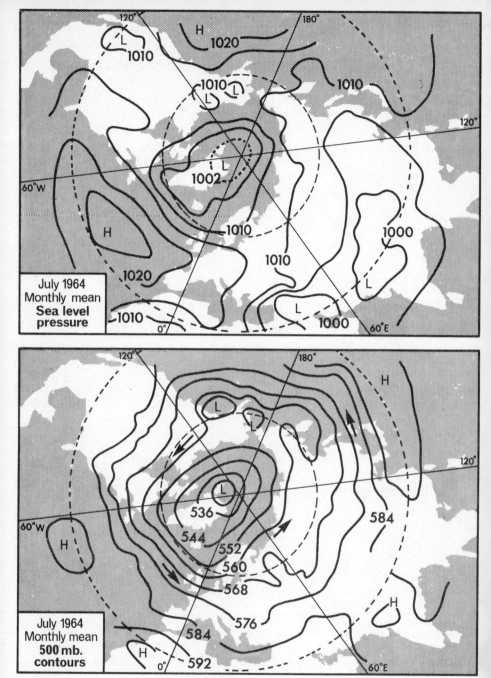

July 1964
Monthly mean
**Sea level
pressure**

July 1964
Monthly mean
**500 mb.
contours**

striking at the highest level shown, 100 mb. In winter the circulation at
100 mb is stronger in higher latitudes than it is at the lower level 300 mb,
whereas in summer there is little gradient and the winds are light. The
height of 100 mb in middle and high latitudes is well into the stratosphere

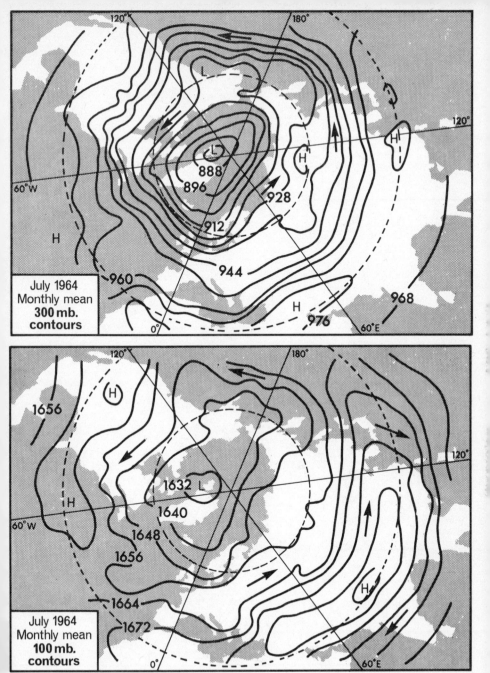

July 1964
Monthly mean
300 mb.
contours

July 1964
Monthly mean
100 mb.
contours

where summer winds decrease with height. At a still greater height the
summer winds are from the east; that is to say the direction of the
circumpolar vortex changes completely from winter to summer, but at
100 mb and below westerlies are the rule all the year round.

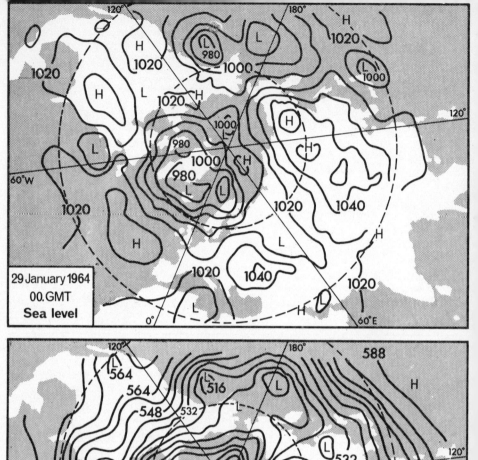

29 January 1964
00. GMT
Sea level

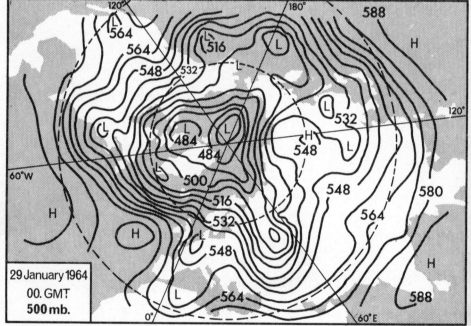

29 January 1964
00. GMT
500 mb.

19. *Notes on Charts for Midnight, 28–29 January 1964.*

Compared with the previous situation for 13–14 January, the patterns over the Atlantic sector are now very much back to normal with low pressure over and to the north of Iceland (the 'Icelandic Low') and a large anticyclone in mid-Atlantic (the 'Azores High'). There is however a lingering tendency for the upper contours, which are tightly bunched across the Atlantic (where there is a strong jet-stream), to fan out and begin to meander when they reach Europe with the formation of cut-off centres – over the Mediterranean and over Russia. Something similar often occurs over the east Pacific where tightly packed contours corresponding with a strong westerly wind across the ocean frequently fan out on aproaching the North American continent, some going northwards across Alaska and others far south to California, with a corresponding fanning or splitting of the wind current.

Once more it is possible to count some 20 centres of high or low pressure on the surface chart and there may be an equal number of minor centres not shown. The concept which treats the disturbances as 'turbulence' in the general circulation of the atmosphere seems very appropriate when one inspects a series of charts for the whole hemisphere in this way.

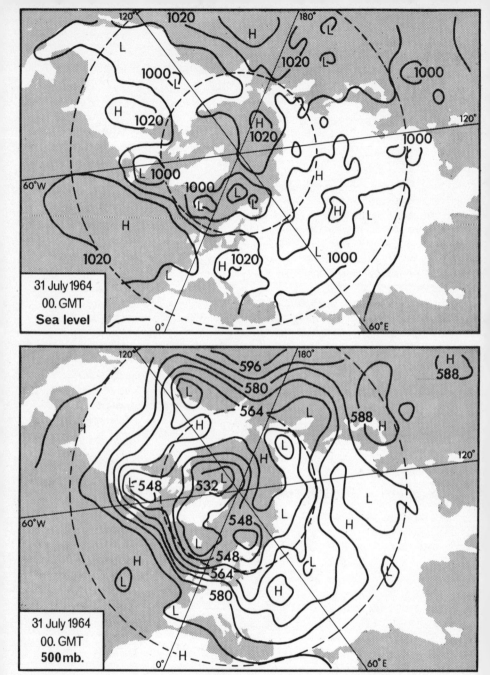

31 July 1964
00. GMT
Sea level

31 July 1964
00. GMT
500 mb.

20. *Notes on Charts for Midnight, 30–31 July 1964.*

This case is chosen as an example of a summer situation for comparison with the winter situations of the preceding January illustrated in Plate 19. Both the sea level and the 500 mb charts are similar to the winter charts with numerous centres of low and high pressure at the surface and, broadly speaking, a circumpolar circulation of westerly winds. Across the northern part of the Atlantic the gradients are strong and south of Greenland there is a developing depression with an open warm sector. This was an unsettled month in north-west Europe and across the Atlantic.

The most striking difference from the winter circulation is that the upper circulation at 500 mb is now mostly north of 45° N. In other words the circumpolar vortex is much shrunken in size and in almost all longitudes the 500 mb winds are light south of 45° N. This is a regular climatological feature corresponding with the northward migration of the sub-tropical anticyclones and other climatic belts, in harmony with the apparent movement of the sun.

One notes the large low pressure to the north of India. This is the summer monsoon low but these charts are not sufficiently detailed to show the tropical circulations.

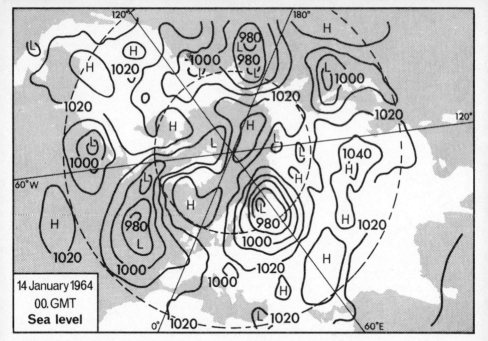

14 January 1964
00. GMT
Sea level

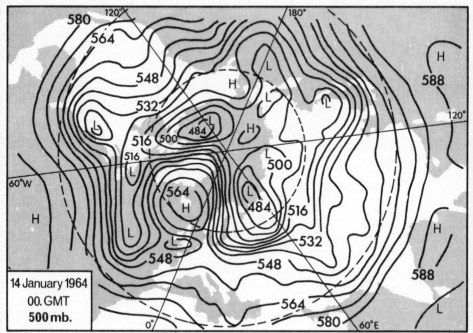

14 January 1964
00. GMT
500 mb.

21. *Notes on Charts for Midnight, 13–14 January 1964.*

As is usual in a winter situation, the Siberian anticyclone and the Aleutian low are well in evidence on the sea level chart. Pressure is also generally high over North America, another normal feature, but over the Atlantic the situation is quite abnormal. There is a large anticyclone centred over Iceland where on climatological averages there should be a depression, while pressure is low where the so-called 'Azores High' is normally located. This reversal of the normal pattern again constitutes a block, as discussed in the Notes for 3–4 January, although this time it is over the Atlantic rather than Europe. The upper contours shown on the 500 mb chart again follow a very large-scale meander with easterly winds over a large part of the Atlantic north of about 50° N. Most of the contours in this case show an S-shape when viewed from the South towards the North, whereas on 3–4 January the shape was a reversed-S. Looking at the pattern in the direction from Europe towards Greenland we may recognize the shape of the Greek capital Ω and the pattern has been called an Omega-block, but these descriptive labels are not particularly helpful.

Except over the Atlantic-Europe sector the situations for 3–4 January and 13–14 January have an obvious family likeness. This illustrates the well-known fact that this particular sector is very liable to variations in weather patterns on the large scale.

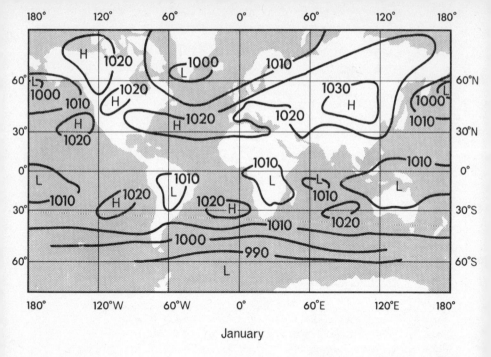

January

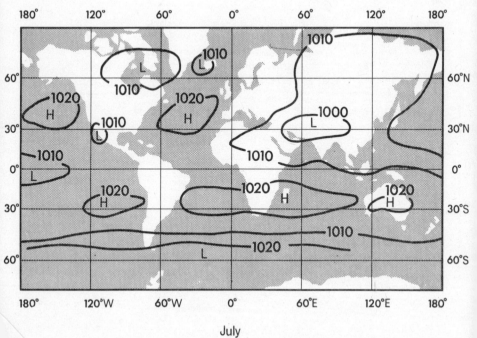

July

22. Average pressures in mb at mean sea level for January and July.

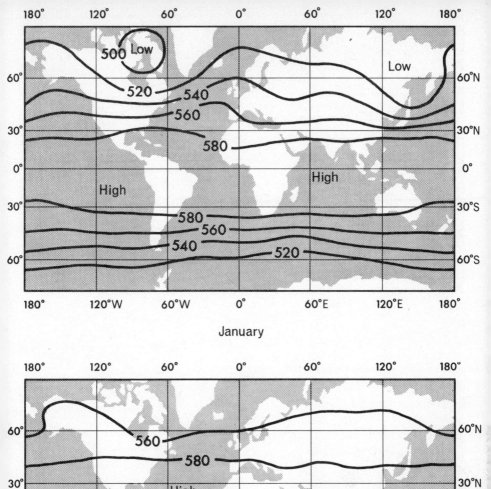

January

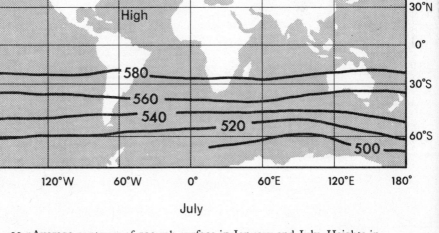

July

23. *Average contours of 500 mb surface in January and July. Heights in hectometres.

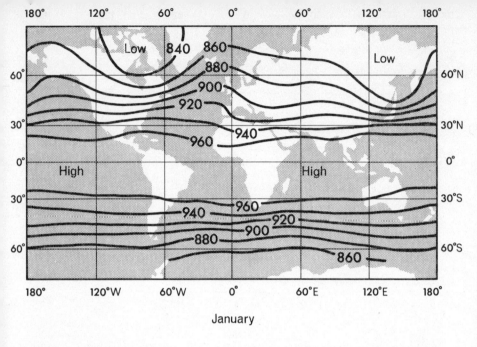

January

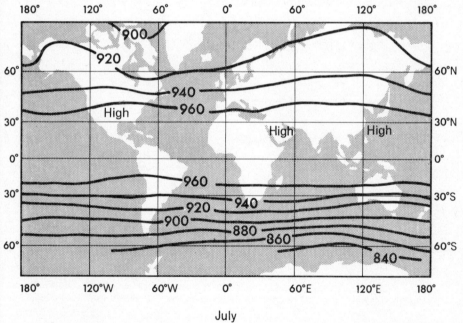

July

24. Average contours of 300 mb surface in January and July. Heights in hectometres.

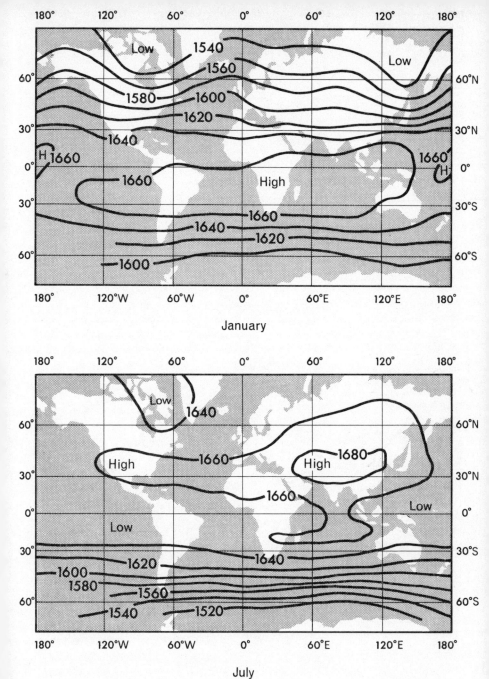

25. Average contours of 100 mb surface in January and July. Heights in hectometers.

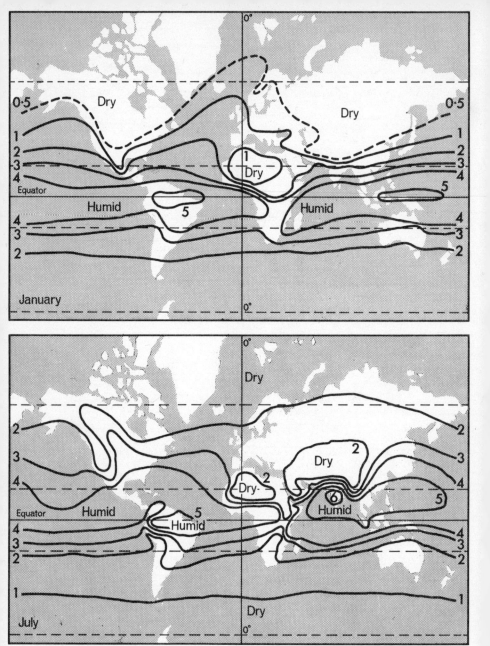

26. Total mean moisture in atmosphere in January and July. Equivalent depth of water in centimetres.

technique a weather situation, say at 6 a.m. on a particular day, is represented by a surface map, giving amongst other things the mean sea-level isobars, and a set of upper-air charts for levels well into the stratosphere. Observations of winds, temperatures, etc., appropriate to the several levels are plotted at the correct geographical positions on the charts to fill in the picture; a three-dimensional picture and also a moving picture, for winds and weather are for ever changing and the task of the synoptician is to study these changes.

The problem is one of fluid mechanics and the air motion or wind is of primary interest, being represented on the maps by actual observed data but also to a useful approximation by the run of the isobars or contour lines, for large-scale winds follow the direction of these lines rather closely. The high values of pressure or contour height are to the right of the direction of flow in the northern hemisphere, to the left in the southern. Thus westerly winds near the surface require pressure decreasing towards the poles in both hemispheres and westerly winds at a higher level require that the contour height of the pressure surface should fall polewards. This law of the winds is readily derived from the general laws of motion and is a consequence of the fact that the large-scale wind currents moving over the face of the earth do so rather steadily with little acceleration. The motion is therefore, roughly speaking, under balanced forces requiring a flow parallel to the isobars or contours. The same cannot be said of the local winds experienced on the coasts or in valleys or in the lee of hills, nor of those which blow through the streets or gardens of built-up areas: these are controlled by local temperature and topographical characteristics, and give but poor indication of the large wind currents affecting the country as a whole.

The physical state and motion of the atmosphere and their changes with time as revealed by the sequence of fully analysed synoptic weather-maps are the essence of meteorology, and to describe all the events encountered in synoptic studies is to cover the greater part of the entire science, observational and theoretical. In this sense, synoptic meteorology, comprehending not only the techniques but also the description and the theory, is almost the whole of meteorology and certainly every chapter in this book deals with matter with which the complete synoptic meteorologist must be familiar. On the other hand there are aspects of the science which are revealed only by the synoptic method and which are therefore regarded as peculiarly the province of the synoptic meteorologist. A brief description of some of these provides the material for the next chapter.

Chapter 9

Large-scale Weather Systems

THE method of analysis which we call the synoptic method, being based on the study of series of synoptic weather-maps, holds a supreme position in the science of meteorology as the technique which reveals the structure of the all-important weather-systems, depressions, anticyclones, and other systems of large scale, even on the scale of the world as a whole. Historically, synoptic charts have grown from small beginnings, when it was possible to collect data quickly only from near-by national stations, to the maps at least for a hemisphere if not for the whole world which are now constructed at every major national centre. For many years, also, synoptic charts were limited to observations made exclusively on the earth's surface and it was not until recent years that upper-air charts could usefully and regularly be constructed. In this way our knowledge of the structure of the world's weather has been built up from the small national scale to the large world scale, and from events near the ground to events in three dimensions. As a legacy of this history we still give primary attention to the surface weather-map for our own part of the world, especially when providing, through radio or newspaper, information for the general public. For our present purpose, however, we may more conveniently begin by taking a synoptic view of the atmosphere as a whole.

As the two hemispheres behave to a large extent independently, it is convenient to restrict attention to the northern hemisphere. At the same time, instead of the traditional synoptic charts for the earth's surface, which are excessively complicated by a score or more depressions and anticyclones in the hemisphere, we look first at the patterns revealed by charts for the middle or upper troposphere, for levels where the pressure is 500 mb or 300 mb in the heart of the upper winds. The circumpolar chart, with the North Pole in the middle surrounded by the circles of latitude, is well suited for the purpose. A number of examples are given in Plates 17–21 with explanatory captions.

Perhaps the first most striking features of a series of contour-maps of 500 mb and 300 mb, and indeed of any pressure levels from about 700 mb to 200 mb, are that the patterns at all levels tend to be very similar in shape and geographical position and that the contour lines take the form of more or less distorted circles centred near the Pole with the lowest values near the Pole and the highest values in the tropics. From the relationship between wind and pressure distribution these contour patterns imply upper winds roughly in the form of a circumpolar circulation, often called the circumpolar vortex, the direction of flow being broadly from west to east – westerly winds in other words.

Again making a generalization, it is noticed that the contour gradient, and the wind speed which is closely related, increases upwards and reaches a maximum near the tropopause which is found not far from 300 mb (30,000 feet) in middle latitudes although much higher, near 100 mb (55,000 feet), in the tropics. With further height above the tropopause the winds usually decrease in strength. The circumpolar west wind vortex also tends to reach its greatest strength somewhere in middle latitudes and combining the two together, the variation with height and the variation with latitude, a core of maximum westerly wind strength is to be found somewhere in middle latitudes and near the tropopause. The maximum can generally be traced all around the earth, more or less continuously, and is known as the jet stream. It is a good graphic name coined in the United States in 1946 or thereabouts when this feature of world winds was first inferred but it is not a very good name if it leads people to think that there is a special kind of current called a jet stream which exists independently in the upper air. The jet stream is the central core of the circumpolar westerly winds, where the speed often exceeds 100 knots and sometimes 200 knots, and it is sometimes a remarkably sharp maximum with the speed falling away to half the strength within 200 or 300 miles, but it is the centre of a broad stream, so to speak, not a local current.

The simple concept of the circumpolar westerlies with its jet stream maximum is, moreover, a generalization which is sometimes hard to stretch to cover the complexities of nature. Canadian meteorologists, for example, who must make regular detailed studies of the winds between the coldest Arctic regions and the subtropics, have long argued that three maxima in three different latitudes, three jet streams in other words, fit the facts much better than does the simple model and, more generally, in almost all longitudes it is usually necessary to distinguish between two quite separate jet streams within the broad circumpolar westerlies. The first occurs in middle latitudes at a height of near 300 mb (about 30,000 feet) and is a very variable feature bound up with the depressions and fronts of these latitudes whereas the second is much higher, near 150 mb (about 45,000 feet), and at a much lower latitude within the subtropics where typically the weather is fine and the barometric pressure is high – the subtropical anticyclones.

However, with all its complications, the idea of a circumpolar vortex of westerly upper winds is a useful beginning and serves as a background against which to describe and to classify the variations of pattern as they occur. Some of these variations may now be briefly described.

In the first place, the general strength of the circumpolar vortex varies over periods of a week or two and at any one time may be stronger or weaker than average. It is also very much stronger on the whole in winter than in summer and the strength may be represented by some suitably defined index, a west-wind index or 'zonal index' which measures wind speed over some convenient latitudinal belt, averaged all round the earth or over limited sectors of longitude. Then the centre of the vortex may be eccentric to the pole or may split into two or more distinct centres. More often than not in the winter time there are two centres in high latitudes, typically over the main land masses of Canada and Siberia, but these are subject to large excursions to other longitudes from time to time. All these features are illustrated in the charts of Plates 17–21.

Perhaps the characteristic of the upper westerlies which has received most attention is its meandering property. Instead of running in one smooth circle round the polar centre the contour lines invariably exhibit a waviness with more or less pronounced displacements towards the pole (in so-called ridges, for here the pressure is relatively high for the latitude and the circulation is anticyclonically curved) or towards the equator (in so-called troughs where pressure is correspondingly low and the flow cyclonic). By measuring the distance in the west to east direction between successive troughs or successive ridges a 'wave length' can be defined, or by counting the number of waves in the whole way round the earth we may speak of a 'wave number'. It was C.-G. Rossby who, some twenty-odd years ago, popularized this mode of thinking about the planetary circulation and made important contributions to the theoretical explanations of the existence and behaviour of the large-scale features which are consequently often called Rossby waves, although more recent dynamical work makes one doubt whether the wavy meanderings which actually occur are basically of the simple kind which Rossby first discussed. In recent years, the description of the upper circulation in terms of wave motion has become very fashionable and on the whole illuminating. By harmonic analysis, by filtering, or smoothing in various ways the irregular meandering pattern may be reduced to some sort of mathematical order, but even without these technical procedures it is obvious at a glance that the circulation has large-scale features showing perhaps two or three major waves around the earth with smaller scale variations superposed upon them. In some contexts, it seems helpful to describe the 'waves' as 'turbulence' within the general circulation of the whole atmosphere, and the small-scale waves as turbulence within the long waves; and the very fact that the atmospheric motions are sometimes called waves and sometimes turbulence is evidence of their complicated nature.

To continue with the description of typical modes of behaviour, it may be noted that the large-scale troughs and ridges in the meandering circumpolar circulation vary greatly in amplitude over periods of a few days and from time to time may completely disrupt, throwing off detached closed circulations on either the polar or the tropical side. The cut-off circulation on the polar side is anticyclonic and on the equatorial side cyclonic, and in either case once detached from the main westerlies may persist for several days in favoured geographical positions with important consequences for the weather.

A pattern characterized by cut-off cells in the circumpolar westerly winds is called a blocked pattern, the emphasis here being on the fact that the westerly upper wind is divided into two parts, the one being displaced northwards, the other southwards, leaving the middle zone as a region of little or no westerly wind: in some cases there may be an extensive zone of easterlies. There are geographical regions which are strongly preferred for blocking patterns, the one being over the east Atlantic and west Europe to the Polar seas, the other rather similarly located over the east Pacific. Since stagnation in the upper winds generally means stagnation in the progression of weather the blocked upper-air pattern is associated with persistent spells of weather including the long spells of easterly winds which in winter bring the coldest weather to the western parts of the temperate continents.

The impression one gets from examining long series of upper-air charts, is of a swirling circumpolar circulation with distortions and cut-off eddies such as might develop, although much more rapidly, within a flowing river and owing nothing to the thermal processes, the conversions of heat and energy, which we know are essential to maintain the circulation of the atmosphere. If, indeed, we could take the westerlies as a starting point, as being created and maintained by some unspecified mechanism, it is certain that long-wave disturbances would either develop spontaneously by inertial instability or be initiated by the large land masses and mountain barriers which obstruct the free eastward movement. It has been abundantly proved by dynamical calculation that starting with the flow as it actually exists at any time, say at 6 a.m. one morning, the flow twenty-four hours later, at the middle level of about 500 mb pressure, can be predicted surprisingly accurately by treating the atmosphere at that level as though it were a uniform fluid, like river water. But in spite of this property which makes it, up to a point, both interesting and profitable to study the upper westerlies in their own right, it must never be forgotten that the atmosphere is far from being such a uniform fluid. It has large temperature differences, especially between high and low latitudes, between polar and tropical air, it is baroclinic not barotropic to use the technical terms, there are large variations of wind with height in speed and also in direction, there are extremely important vertical motions associated with anticyclones and depressions, and there is the all-important water cycle of evaporation, condensation, and precipitation. All these aspects of weather processes, taken in conjunction with the inertial movements of the westerlies, go to form a three-dimensional complex impossible to describe completely in other than mathematical language but containing characteristic features which repeat themselves more or less faithfully time after time.

The cyclonic depression is the large-scale weather system which has received the greatest amount of attention throughout the century of synoptic weather-forecasting, and for the very good reason that it is within the circulation of depressions that we experience most of our strong winds or gales and by far the majority of our heavily-clouded skies and precipitation, both rain and snow. The close association between disturbed weather and the pressure distribution at the earth's surface is the explanation of the usefulness of the barometer as a 'weather glass'; it was for generations the basis of weather-forecasting, almost, one is tempted to say, the sole basis, and the accuracy which can be attained in predicting the developing, intensity, configuration, and movement of depressions remains today the main limitation upon the accuracy of short-range forecasting.

For a long time, indeed until less than twenty years ago, a satisfactory explanation of the fluid dynamics of the depression eluded meteorologists, mainly because upper-air observations were inadequate for calculation, but in recent years the difficulties have mostly been resolved.

The depression, or cyclone, is, by definition, a region of relatively low barometric pressure at the earth's surface. When a new centre develops the pressure falls for a day or so over an area some hundreds of miles across and the wind, responding to the pressure gradient, picks up its cyclonic circulation (in the Northern hemisphere anticlockwise, in both hemispheres anti-sundialwise). This circulation or spin of air in the same direction as the earth's rotation appears to emerge from nowhere, in defiance of the laws of angular momentum, but the appearance is merely an earthbound illusion. A spectator out in space would see the earth's atmosphere spinning with the earth, round the poles once every day; a very rapid spin for such a large body, requiring a speed of 1000 miles per hour near the equator. Against this general background, the circulation near the earth's surface in a cyclonic depression would appear as a smaller vortex embedded within the whole, and spinning somewhat more rapidly in the same general direction. The correct explanation is then readily to hand, for the extra rotation of the cyclone would be provided for by a contraction in the area covered by the air of the cyclone: a necessary and sufficient condition for the development of a cyclone is a contraction in area by the convergence of the air or inflow towards a centre. The inflow implies vertical upward motion, which of course readily explains – in a general way – the cloudiness and rainfall. All that remains is to find a mechanism which will remove air laterally at a higher level or, in other words, allow it to diverge from the centre in order to balance the pressure at the earth's surface. In the event the upper divergence must be somewhat greater than the lower convergence for there is a fall of pressure in the cyclone centre but this is, rather surprisingly, not a point of great dynamical significance. A fall of pressure of 50 mb is remarkable but requires an average divergence in the air column above of only 5 per cent (50 mb in 1000 mb), whereas the corresponding convergence of air, or contraction of area, required to produce the spin of the winds from the earth's rotation may be more like 50 per cent: the problem is to provide for a divergence of like amount in the upper air. Such a divergence should, in reciprocal fashion, produce a clockwise or anticyclonic circulation but depressions of the kind we have in mind do not have anticyclonic circulations in the upper troposphere and there is evidently some error in the argument. Actually the error arises from an unsatisfactory formulation of the problem, that is from trying to understand a depression as though it were formed in an originally stationary atmosphere, whereas it is actually formed in a region of the circumpolar westerlies where the winds increase very markedly with height. It is this shear of wind which, in rather a subtle way, allows the divergence in the upper wind to occur in the same region as the convergence in the lower wind as is required for the development of the cyclonic depression. At the same time, in some neighbouring longitude, there is likely to be divergence near the ground and convergence in the upper air coupled together by a sinking motion as is typical of the development of the anticyclonic high pressure. Dynamical theory has in this way not only provided a fundamental explanation of cyclone development but has provided at the same time an explanation of anticyclone development, so attaining the hall-mark of good theory, that it explains more than it set out to explain.

At about this stage in the argument the dynamicist would want to have done with qualitative arguments and get down to mathematical calculations for it is only by calculation that we may become reasonably sure that the arguments are valid, but here we may take the mathematics on trust and continue the explanation with the aid of schematic diagrams. We imagine as a beginning the belt of more or less westerlies of middle latitude, the precise direction is not important, represented by the first diagram, Fig. 8 (a). The air is cold to the north and warm to the south and, as a consequence, the winds, in this baroclinic zone increase with height to a maximum at the base of the stratosphere. Now let us suppose that by some disturbing mechanism the whole current is twisted cyclonically and anticyclonically as indicated in Fig. 8 (b): the result is a wave-like disturbance embedded in the stream and carried along with the average speed of the current which will be the speed at a middle height. Above that height, in the upper troposphere, the air moves more quickly through the wave form and must pick up the cyclonic curvature in moving into the trough of the wave marked C and again lose this curvature and pick up the anticyclonic circulation to round the ridge marked A further east. At a low level, especially in the lighter westerlies near the surface of the earth, the circumstances are, however, essentially different in that the air does not flow through the pattern but the wave pattern itself is carried forward through the air so that between C and A the lower air is gaining cyclonic spin. At the higher level between C and A there must be horizontal divergence to create anticyclonic spin; at the lower level in the same region there must be horizontal convergence to produce the cyclonic spin. In this way, owing to the increase of wind with height, the balancing patterns which we are looking for are found to arise naturally in any wave-like distortion of the general flow. The whole baroclinic zone is therefore essentially unstable and will twist up into self-sustaining cyclones and anticyclones at the smallest provocation; and that is what does happen in nature. Calculation shows that

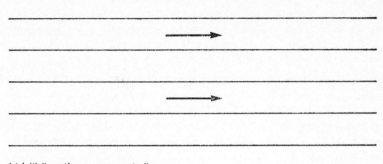

(a) Initially uniform upper westerlies

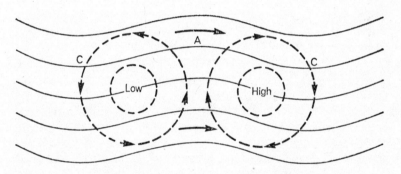

(b) After formation of disturbances. Broken lines indicate surface isobars

Fig. 8. Schematic illustration of development in the westerlies. The twisting of the cyclonic and anticyclonic circulations around the centres marked LOW and HIGH produces a wavy distortion in the upper stream lines. This requires upper cyclonic spin in the trough marked C and anticyclonic spin in the ridge marked A. The air flowing through the pattern must therefore diverge between C and A and converge between A and C, so automatically providing for the fall of pressure in the Low and the rise of pressure in the High. The system is therefore unstable and cyclones and anticyclones develop spontaneously.

certain conditions must be satisfied: the whole disturbance must not be too large, not greater than about 1000 miles across more or less according to the strength and breadth of the upper winds, as otherwise the waves are prevented from developing by an effect of the earth's curvature and no spontaneous developments will occur. Again the disturbance must not be too small if it is not to be inhibited by the vertical stability of the atmosphere and calculation shows that between the two limits there is just enough latitude to accommodate the kind of depressions and anticyclones with which we are familiar.

The theory developed in this way would suggest that the zone of the baroclinic westerlies would be perpetually unstable, continually giving birth to new disturbances, and this is true. It would also suggest that the stronger the thermal contrast the more vigorous would be the disturbances and this is also true. But it would also suggest that any part of the circumpolar westerlies would be equally likely to give rise to disturbances if they form spontaneously as the outcome of dynamical instability: and this would not be true. On the real earth, in the real atmosphere, as distinct from the mathematician's simpler model atmosphere, there are always preferred regions, perhaps in the lee of mountains or the western sides of the oceans, where flow patterns are particularly likely to set up cyclonic disturbances, just as there are other preferred regions, perhaps on the windward side of mountains or the eastern sides of the oceans where flow patterns are particularly liable to be distorted into the anticyclonic mode. Considerations of this kind open up a big subject, indeed the whole problem of climatology as studied from the synoptic and geographical points of view, and it would be unprofitable to attempt further clarification in brief compass.

The theory has indicated how baroclinic depressions and anticyclones form and develop but has not given the full life-cycle from development through maturity to decay and disappearance, equally essential aspects, for the number of cyclones and anticyclones on earth, although varying from time to time, does not change in the long run so that, on average, as many systems vanish from the face of the earth as are created in any length of time. The problem of the disappearance of depressions and anticyclones does not, however, lend itself to attractive formal solutions as there are so many complications in nature. Many, perhaps the majority, lose their identity at some stage by becoming absorbed in the circulation of a larger or more vigorous system; many, and far more than is generally realized, continue in existence as surface pressure features for many days or weeks, often becoming regenerated as they trigger off a new baroclinic disturbance; but, broadly speaking, the process which is continuously tending to destroy circulation systems is friction between the atmosphere and the earth below. Acting alone, friction is sufficient to reduce a system to half its intensity in two or three days and, on the long time-scale, all disturbances are doomed to be lost without trace.

Omitting then the ultimate process of decay and disappearance, the typical life story of the baroclinic depression or baroclinic anticyclone may be represented by the series of schematic diagrams in Fig. 9. As the developments take place in regions with marked horizontal thermal gradients and the winds increase generally with height through the troposphere, the model is three-dimensional in a way which is difficult to illustrate in simple drawings, but a sufficient indication is obtained from the two series of diagrams, referring respectively to the cyclonic and anticyclonic case. In each diagram is shown the surface pressure distribution which indicates the wind circulation near the surface, together with schematic isotherms to indicate the intensity and direction of the thermal gradient, that is from the warm to the cold air. In the cyclonic case, the evolution is from a weak surface circulation in a strongly baroclinic region with only a wave-like distortion of the upper winds to a strong surface circulation in a very distorted thermal pattern, cold over the centre of the low, which implies a cut-off closed circulation throughout the troposphere. The final stage tends, therefore, towards a circular vortex with vertical axis; the general wind over the centre is light at all levels and the system as a whole moves but slowly; often it stagnates completely. The process of low-level convergence and upper divergence which has allowed the cyclonic vortex to develop has necessarily involved upward motion, the air has become generally cloudy and rain will have fallen. In the anticyclonic case, the final stage corresponding to the cut-off cold cyclone is the cut-off warm anticyclone, but in neither case does the development always proceed to completion with the emergence of an identifiable new mature system. Rather more typically, the new centre will lose its identity by merging

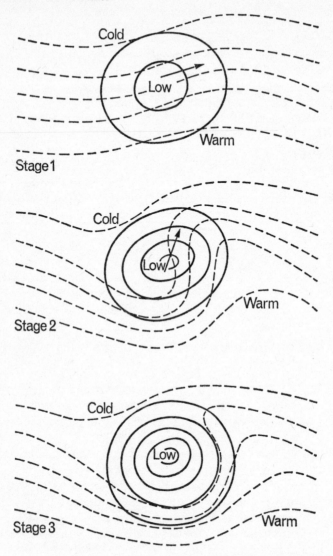

Fig. 9. Schematic diagram illustrating stages of development of depression and anticyclone in a zone of large temperature gradient between cold air to north and warm air to the south.

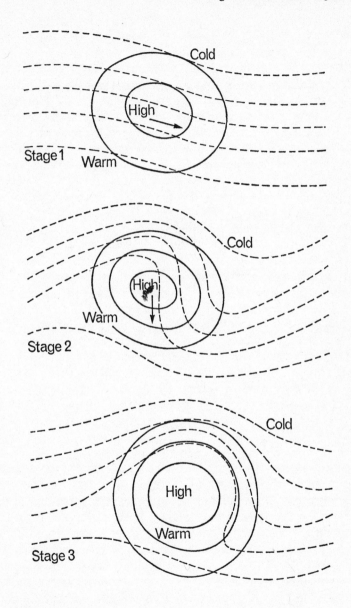

with an established cold depression or warm anticyclone which is thereby maintained. The very persistent lows and highs which dominate the general circulation of the northern hemisphere, sometimes for weeks at a time, for example the Icelandic low and the Azores high, are maintained in this way by absorbing new developments which form in the strongly baroclinic region which occurs between them and farther west. The creation of a new and independent major cyclone or anticyclone which reaches maturity without merging with a pre-existing major centre, heralds a change of weather type which in middle latitudes may commonly occur at intervals of a week or two.

The account just given of the dynamical development of cyclones and anticyclones has been in terms of baroclinic zones, broad belts with strong temperature gradients between the warmer and the colder regions and associated with winds increasing upwards to a maximum in the jet-stream near the tropopause. Where, it may be asked by anyone who has previously studied meteorology, are the fronts and air-masses which have dominated writings on synoptic meteorology for forty years; where is the warm-sector depression and the occlusion process of the Norwegian school which have had an honoured place in every textbook for so long; what of the fronts shown on the weather-maps every day in the newspapers or on television? These are fair questions and the answer is that in recent years it has been proved by reliable calculation that the essential major properties of depressions can be derived without introducing the concept of a front at all and, what is equally notable, that not only depressions but also anticyclones emerge from the same basic dynamical theory, whereas in the language of fronts and air-masses they called for quite separate treatment. The conclusion forced upon us is that fronts are no more essential to the basic theory of cyclones than they are to the theory of anticyclones; but, while theories may be superseded, facts remain facts and since fronts exist as certainly in 1964 as they did in 1924 they must be explained as extremely important secondary phenomena which very commonly appear in depressions but not often in anticyclones and behave more or less in the manner described in the standard texts. This book is being written at a time of transition when general dynamical theories elaborated by a minority of research meteorologists are superseding established ideas of synoptic analysis in the way that is characteristic of advances in scientific theory. The older ideas are not being eliminated but within the wider framework they do call for reformulation to retain their validity and we may take advantage of this interesting stage in the science to discuss fronts and air masses in the new light. It should, however, be admitted that an entirely satisfactory dynamical theory of fronts is not yet available and that until convincing calculations have been carried out we shall remain uneasy about the true explanation of some of the characteristic behaviour of fronts.

In general terms, the frontal and air-mass concepts may be put quite briefly. We are to think of the atmosphere in the troposphere, the seat of weather, as being divided into moving masses of air of varying geographical dimensions up to continental size. Each mass has substantially uniform properties, relatively warm or cold, stable or unstable in the vertical, humid or dry, and is separated from neighbouring masses by narrow belts of transition often sharp enough to be represented by a single boundary line on a weather-map: this boundary line is called a front and the continuation of the sharp boundary through the atmosphere above is a frontal surface sloping in such a way as to allow the warmer air mass to overlie the cooler. Upsliding at the surface is often the cause of clouds and rain.

In the early formulation of the concepts by the Bergen school of meteorologists the main interest lay in middle latitudes, especially the Atlantic and European sector; the two main air masses were of polar and tropical origin; the dividing boundary was called the polar front, and the scheme was applicable to the majority of weather situations without too much forcing. The typical depression was diagnosed as a disturbance forming on the polar front and going through a life-cycle of formation, development, maturity and decay in which the fronts and air masses played the major rôles as described in the following summary account.

In the early stages, the air masses of polar and tropical origin would be lying alongside each other, separated by the polar front or frontal surface in a quiescent condition with little movement. The warm air would be moving more or less from west to east, shearing past the colder air which would be relatively slow-moving

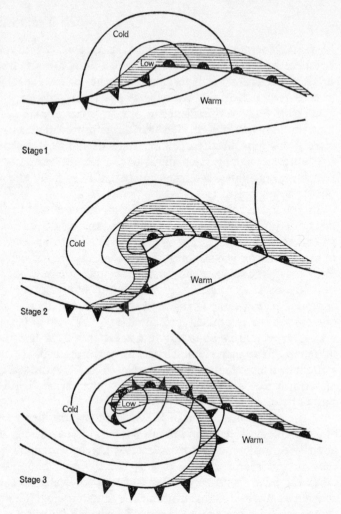

Fig. 10. Cyclonic development according to Bergen school (after Bjerknes)

Stage 1. Initiation. Shallow depression and wave-like distortion of the polar front

Stage 2. Rapid development. Deepening depression and large warm sector

Stage 3. Fully developed deep occluded depression

    Warm front—lines with semicircles

    Cold front—lines with spikes

    Occlusion—lines with alternating spikes and semi-circles

    Rain areas shaded.

as in Stage 1 of Fig. 10. The birth of the depression would be heralded by a slow fall of pressure, over an area some hundreds of kilometres in extent, greatest at the front, with a corresponding circulation of winds and a distortion of the front into a wave-like shape. The front would naturally be carried along by the winds so that the part ahead of the centre, called the warm front, would replace cold air by warm, whereas to the west of the centre the circulation would replace warm air by cold at the cold front. The region between the warm front and the following cold front was called the warm sector. The process having been initiated in this way would be maintained by a continuing fall of pressure or deepening of the depression, continuing increase of winds, possibly reaching gale force, and continuous distortion of the fronts in the manner illustrated in the diagrams, until the cold front gradually overtook the warm front at the so-called 'occlusion' and the depression was then mature. Many of the well-known properties of depressions were fitted rationally into the model, especially the clouds and rainfall. Ahead of the moving depression cirrus cloud first appears, followed by altostratus and nimbostratus with falling rain, the whole system – some hundreds of miles in width and extending outwards from the centre of the depression in the radial direction – being beautifully explained as the effect of the warm air of the warm sector ascending and sliding over the cold air. The cold front and the occlusion were also assigned cloud and precipitation structures as illustrated in Fig. 11. The frontal model of the depression of middle latitudes was extremely helpful, especially in weather-forecasting, as it introduced a new frame of reference for a systematic structure of winds and weather where previously there was little that could be done beyond the drawing of isobars. In the history of synoptic meteorology, the discovery of fronts and the development of air-mass analysis will always retain a distinguished place, linked with the names of the early workers in Bergen, J. Bjerknes, T. Bergeron, and others. The concepts themselves will also find a permanent place in the science, but with rather less prominence than seemed likely twenty years ago. In the generation following the early work, the ideas gradually became accepted all over the world, being adapted to suit the local climatic conditions and elaborated until textbooks were written listing dozens of different 'air masses' and every belt of cloud and

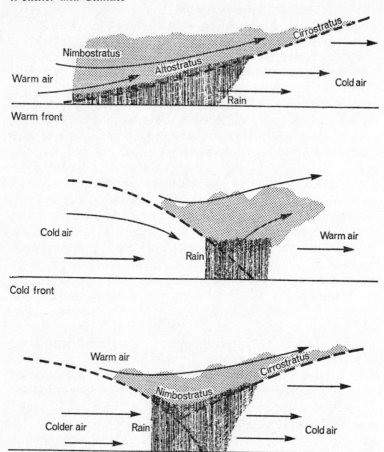

Fig. 11. Schematical vertical sections through fronts according
to Bergen School.

rain was assigned to some kind of front. Of course, in a way, the
atmosphere is like that. There are, indeed, all kinds of minor
differences between one wind current and another, and minor rain
areas are often associated with such differences so that air-mass
analysis and frontal analysis become a matter of taste with criteria
almost impossible to define. Nevertheless, the broad outlines
stand, large differences of temperature and humidity are properly
assigned to differences in the origin and recent history of the air,

while the fronts dividing them are the seat of much cloud and rainfall.

The one factor missing from the air mass and frontal theory was a quantitative dynamical explanation of the behaviour. Why should a frontal surface distort in this way and give rise to a cyclone? Why should the warm sector air ride over the cold air at the warm front and produce clouds and rain? Why should the system occlude? No satisfactory theory was ever produced. It is, however, most instructive to compare the frontal model illustrated in Fig. 10 with the corresponding model of cyclone development in a continuous baroclinic fluid as illustrated in Fig. 9. They have much in common, and are readily combined into one model by inserting fronts in the appropriate regions of the continuous fluid. The cross-section of the warm front needs, however, to undergo some modification to satisfy dynamical theory and observations at higher levels in the troposphere. In Fig. 12 are sketches illustrating the typical warm front according to the two

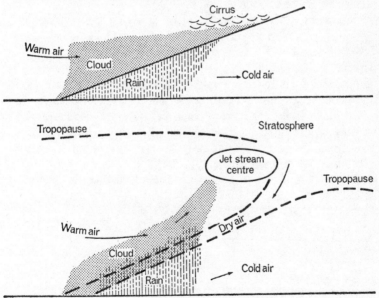

Fig. 12. Sections through a warm front. *Top* conventional drawing; *bottom* drawing based on recent observations.

view points and there are some critical differences. The jet-stream core of strong winds centred below the tropopause of the warm air and vertically above the position where the frontal surface is found at about 5 kilometres height; the sudden decrease in height of the tropopause on the cold side of the jet; the tendency for the cloud mass to separate from the frontal surface and for the upper levels of the frontal surface to be very dry rather than wet: none of these properties is found in the old frontal models. Actually we must not claim too much for modern dynamics for very few calculations have been carried out concerning the behaviour of the frontal jet-stream complex, and not one is entirely realistic. At the time of writing it is known that the British Meteorological Office is working on frontal dynamics and is planning to use a most advanced electronic computer, that is the ATLAS built by Ferranti Ltd. at a cost in the region of £3 million. Its speed and capacity are some ten times greater than those of the computer in use for routine forecasting, and this may serve as an indication of the true complexity of the behaviour of fronts.

The chapter may be concluded by studying the series of circumpolar charts given in Plates 17–21 with explanatory notes. On these small-scale charts it is impossible to show the details of weather associated with the depressions and fronts, but examples of simplified charts are always available in newspapers or in television programmes. Any reader who is interested in further study may obtain for a very small subscription copies of daily weather reports giving very useful weather-maps; these are available from the official weather-service in all major countries. The study of numerous synoptic surface and upper-air charts affords a kind of clinical experience for which there is absolutely no substitute.

Chapter 10

Weather Forecasting by Calculation

THE sciences of the world around us, sometimes called the environmental sciences, include astronomy, geology, geophysics, oceanography, hydrology, and the like, as well as meteorology. It is a rather curious fact that, amongst them all, meteorology should be distinguished by being identified in the public mind with the art of looking into the future, with prediction or forecasting, and to such an extent that if one confesses to being a meteorologist it is as much as to invite the request for an opinion on the coming weather. This is not wholly to the liking of every meteorologist for many among their number, indeed most university teachers and research workers, have never given a forecast in their lives and have no ambition to do so, but, it is nevertheless true that the ability to forecast the weather has been the great bread-winner for meteorology, has directly or indirectly provided funds for the development of the science all over the world, in universities as elsewhere, and continues to do so. The truth is that we cannot as yet control the weather and, in spite of the innumerable ways in which the atmosphere affects our way of life, the economically most useful application of our knowledge has lain in predicting its behaviour. Taking a world total, something like £500 million is spent annually on meteorology and by far the greater part is devoted to meeting the exigent demands of aviation and other interests for accurate information on current and future weather.

While, however, meteorology has developed steadily over the years and at an accelerating rate, the task of putting the practice of prediction on to a firm scientific basis has not proved easy. The forecaster has become more and more knowledgeable about the physics and dynamics of the atmosphere, but when confronted with the very specific problem, to say what tomorrow's weather may be, instead of making, as one might expect, methodical calculations to reach a definite conclusion according to the best available theory, he has tended to think of all the processes taking place, of all the possibilities which might arise, and then to rely upon his experience and judgement. Weather-forecasting, rather like medical practice, has until recent years developed as a scientific art rather than as an objective technique; and forecasts, like medical diagnoses, have been the opinions of experts rather than the deductions from established laws. This, one should hasten to add, does not mean that forecasting has been an occult art of little intellectual or scientific interest, for a little reflection will remind us that most of the critical predictions which men are called upon to make, in engineering or agriculture, in business, economics, war, politics, or human affairs, still often depend on the wisdom and judgement of well-informed and experienced specialists, professional experts rather than calculating machines, and we do not value the experts the less on that account. The object of science in all these many fields may even be defined as the substitution of reliable general rules, or laws, and exact calculations for the judgements of wise men and, as would be expected, the more complicated the problem the more difficult it is to attain the goal and put the wise men out of business. Weather-forecasting has been and still is a very complicated problem in physics, and the art of forecasting must be learned, as any other art, by long experience with its successes and failures; but some progress has been made in recent years towards replacing the uncertain art by reliable science, the fallible judgement by the numerical calculation, and this I shall attempt to describe. One cannot however resist a sigh of regret. There is tremendous intellectual satisfaction for the research scientist who originates a systematic method of calculation superior to the judgement of the established expert, but the passing of the expert in favour of the electronic computer means the elimination of a rewarding way of life. The replacement of the expert judgement of the distinguished specialist by a more reliable system of standardized calculation, is the parallel in intellectual life of the replacement of the individual skill of the artisan by the more efficient methods of mass production – and may be regretted for much the same reason.

If one only thinks for a moment of the variations of cloud and sunshine on a showery day, of the endless variations of visibility when motoring on a foggy night, of the fluctuating winds near a coastline or in hilly country, one will readily concede that even to state in full detail what the weather is everywhere is an impossible goal: to determine what it will be at every moment and in every place at some time in the future is surely an inconceivable attainment. Forecasting must then be content with some degree of generalization, and forecasting by calculation has so far been restricted to the large-scale features, to the main wind-systems of the upper atmosphere and the developments and movements of depressions, anticyclones, and other synoptic systems. For this purpose it will suffice to know what is the wind, the pressure, and the temperature everywhere in the atmosphere. These are the quantities which are measured by balloon observations in the free atmosphere, and the problem is to find a method of calculating how these quantities will change as time goes on during the succeeding hours or days, or as long as may be practicable.

Using a rather different language, we may say that weather-forecasting requires the solution of a problem in fluid mechanics, the fluid being compressible and in turbulent motion over the surface of a gravitating sphere. The surface of the sphere is non-uniform, partly water, partly uneven land, and there is friction between air and surface. The motion is thermodynamically active with exchange of energy taking place at both boundaries, with the earth below and with space above, while internal energy transformations are continuous and of several kinds. The fluid is not of constant composition but continuously gains water vapour by evaporation and loses it by precipitation and this process is, incidentally, a dominant source of energy to the system. It is little wonder that many, indeed most, specialist meteorologists could see little prospect of reducing the behaviour to orderly calculation.

Up to the outbreak of World War II it is doubtful if anyone in the world was expecting that in the foreseeable future, calculation by the methods of mathematical physics would begin to take the place of expert judgement in practical day-to-day forecasting. There were of course scientists, of whom V. Bjerknes (1862–1951), the great Norwegian meteorologist and fluid dynamicist, was the most eminent, who expressed a philosophical belief in the ultimate success of the true scientific method in weather prediction. It was good that the banner of the true scientific faith was kept flying at a time when practical forecasters faced with the real difficulties were sceptics, although in fairness it should be admitted that to be a philosophical optimist is easy if one is content to set no time scale and to wait indefinitely. Perhaps too much credit for far-sightedness is often accorded to early speculators who have imagined success, say, in flights to the moon when their only contribution has been a facile flight of fancy but in our present problem respect must be accorded to the Englishman, L. F. Richardson, who in the nineteen-twenties had shown how the problem could really be tackled by mathematical methods. These were remarkably similar to those successfully used a generation later, but Richardson had abandoned the project, after an early failure, owing to lack of sufficient data and computing facilities.

In looking for suitable laws of nature applicable to the problem it is natural to begin with the mechanical law of motion, Newton's Law, which says that momentum is conserved except in so far as it is changed by the accelerations produced by the forces acting upon the substance. In fluid motion this law leads to a complicated mathematical equation containing many terms and it may be helpful to consider a much simpler kind of motion in order to prepare the way. If a heavy body is falling freely towards the ground the only forces acting upon it are the force of gravity and the air resistance and, if we ignore the air resistance, as it is legitimate to do if the body is sufficiently dense and heavy, we have the simplest case of motion under one force: the downward acceleration is constant and equal to the acceleration of gravity, a very simple equation. Even if we consider the astronomical motions of all the planets in the solar system, the only forces of consequence are the gravitational attractions between each pair and between each of them and the sun. This is a sufficiently complicated system but there are only a few bodies and a few forces so that astronomers may calculate the accelerations very accurately and compute the future movements precisely. In meteorology by contrast there is, in addition to gravity, the force of fluid pressure and the force of fluid friction or viscosity as well as an apparent force, a kind of centrifugal force, which arises from the spin of the earth round its polar axis, one rotation each day. Ordinarily one does not notice that one is standing on a spinning sphere and of course the ancients refused to believe in any such apparently absurd notion; there still, I believe, exists a Flat Earth Society whose members adhere to the ancient faith; but the air moving in the great wind-currents for thousands of miles must respond to the motion of the spinning sphere by tending to turn in the opposite direction relative to the earth, so adding another component to the acceleration to which the name Coriolis acceleration has been given – after a French scientist of the early nineteenth century. The law of motion applied to the atmosphere then equates the acceleration of every portion of the atmosphere to the combined effect of gravity, fluid pressure, friction, and the Coriolis acceleration acting generally in different directions and being different for every portion of fluid. It is complicated, but can be written down quite methodically and finds its place in every standard textbook.

That which we choose as our second law may seem a very obvious one, or a very trivial one, but it is a basic law of nature, of classical physics, which was not always obvious and which can be easily overlooked. It is the law of conservation of matter which, applied to the atmosphere, says that air is never destroyed. In order to express the law in mathematical form, a necessary step if quantitative calculations are to be made, we may consider a small volume of space containing air of a known density; the law then says that the density will increase or decrease according as more or less air is packed into the small volume by air movement. Thus the law of conservation of matter applied to the air gives us a relationship between density and air motion.

For our third law of fluid motion we have the so-called 'gas law', or 'equation of state', the combination of what in English physics texts are usually called Boyle's and Charles's Laws. The law relates pressure, density, and temperature in a way expressed by the equation* $p = R.T.\rho$. Here, R is a constant depending on the gas in question but the earth's atmosphere (apart from water vapour) is a gas of highly uniform composition and R is therefore uniform. This law is of supreme importance because as the air moves from place to place, especially if it rises or descends, the pressure changes considerably and so either or both, usually both,

* The conventional symbols are used, that is p for pressure, T for absolute temperature and ρ for density.

the temperature and the density must change in accordance with the gas law.

Fourth in the list of laws we must take one of the most powerful and most general laws of nature, the law of the conservation of energy, which, expressed in the form suitable to the fluid motion, should include: the kinetic energy (depending on velocity and density); the internal molecular energy dependent essentially on the temperature; the rate at which work is done on the fluid by the forces; and any sources or losses of heat, mainly the latent heat of evaporation and condensation, and the absorption or emission of radiant heat.

At this stage it may be time to call a halt to this attempt to discuss complicated theoretical physics without mathematics but the exercise was unavoidable if we were to give any indication of the way in which the theoretical meteorologist must try to solve his different problems. These are some of the basic foundation stones upon which all explanations of weather phenomena must be built and there is, as a rule, no avoiding them. It is also a convenient point to stop because we have arrived at four laws which, expressed as four mathematical equations in four variables, pressure, temperature, density, and air motion, are in principle sufficient to solve a wide range of problems given suitable facts of observation. The difficulty lies in finding a practical way of applying the known theoretical laws to the atmosphere with all the complications and incompleteness of observations.

The breakthrough came when, with the introduction of many radio-sonde stations during World War II, it became possible to construct weather-maps showing the patterns of pressure and wind at upper levels some 10,000 or 20,000 feet above the surface. The patterns, as illustrated in Chapter 9, were very characteristic when drawn for large areas of the hemisphere and revealed features hitherto only vaguely recognized. The impression given was that of a broad current of wind roughly blowing round the hemisphere continuously from west to east, but with many large wave-like meanderings between low and high latitudes. The features of the patterns, the poleward crests of the waves and the troughs nearer the equator, the belts of strong winds and other regions of more sluggish flow, moved and varied in shape from day to day but in a coherent way such that one recognizable feature could often be traced for many days. C.-G. Rossby, the Swedish oceanographer and meteorologist working in America, to whose work references have already been made, had shown before the war that this kind of flow had something in common with inertia flow in a liquid of uniform density. If, for example, water in a vessel is given a certain pattern of flow and then left to itself, the patterns will move and distort with the liquid until dissipated by friction. The simplest case to study would be one where the fluid is uniform in density, moving without turbulence or friction and without external gain or loss of energy. In such a moving fluid there is no convection, no significant vertical motion, no change of temperature anywhere and no 'weather' of any kind: the only obvious relevance to weather forecasting is that the fluid motion changes with time just as the winds vary with time. It was Rossby who had already shown that some aspects of the upper winds of the hemisphere could be accounted for on this very thin argument; in fact one could take the flow pattern of the air in the middle levels and treat it as though it were the motion of a very simple fluid of the kind described and obtain predictions which were far from being absurd. The laws of physics for this kind of fluid motion are of course, by comparison with meteorological reality, very simple and the problem can be solved quickly using electronic computing aids. When, taking advantage of one of the very first large electronic computers brought into service at the Institute for Advanced Study at Princeton, it was confirmed that the predictions made of the winds at about 18,000 feet by this method were quite similar to the real thing, Rossby and his colleagues, supported by the eminent mathematical physicist J. von Neumann, were convinced that the first step had been taken and that with further development weather-charts predicted by calculation would compete with experienced forecasters. They have been amply justified.

The calculations even for this simplified 'atmosphere' are lengthy because the dynamical law which controls the motion is the conservation law of vorticity or spin. The calculations are made in a step-by-step manner. Starting from the known winds as observed at a given time, the spin is calculated at a large number of places forming a uniform grid over the area, perhaps most of the northern hemisphere. The number of points must be as small as possible to cut down the work but as large as possible in order to obtain a good representation, and a compromise is reached; there may be two thousand points in the grid. Assuming the spin is conserved but carried along with the wind, it is evidently possible to calculate the new distribution of spin over the whole area after a short time has elapsed, sufficiently short to allow the movement to be treated as steady, say half an hour. At this stage, the arithmetic becomes heavy because it is necessary to work out the new winds all over the area to satisfy the new spin, and this is an inverse problem solved most readily on a machine by a process of trial and error, with successively closer approximations. The trouble is that every adjustment to the wind in one place alters the spin at all neighbouring places and it is necessary to go on adjusting until the fit becomes satisfactory all over the area at the same time. Performed systematically in a computer this can be done but the process of adjusting may lead to millions of little bits of arithmetic each taking the machine only some thousandths of a second but adding up to many minutes. To get a forecast for twenty-four hours ahead, the process must be continued step by step many times, forty-eight times if the time steps are each of half an hour, but the whole task can be made practical and be carried out 'in real time'.

It is now some ten years since the first predictions were made in America by computer, and since that time much progress has been made. Computers have become larger in capacity and much faster in operation, while to approximate more closely to the behaviour of the real atmosphere the meteorologists have introduced greater elaboration and now work with the data from several levels of the atmosphere simultaneously, make allowances for adiabatic changes in vertical motions, introduce the effect of heating of cold air over warm surfaces, take into account the mountains, and allow for friction at the ground which is different over the land from over the sea. However, much work of this kind still remains to be done.

In the United States Weather Bureau the computed charts for the upper air have completely taken the place of the experts' estimates, whereas in the United Kingdom – which on available evidence can claim to have advanced farther than any other competitor outside the USA – the computations are, at the time of writing, still essentially supplementary to the judgement of the responsible forecasters. To the old-fashioned forecaster the procedure is almost incredible. The data in the form of miles of punched tapes received by teleprinter from all parts of the geographical area, are fed directly into the computer which is programmed to recognize the relevant information and to discard the remainder, to estimate the values of the pressure or contour height at all the points of the grid and at all required levels taking account of all the data from near-by stations, to scrutinize data for gross errors and reject the doubtful value, and even to use the value previously forecast if there is no better information. In other words, the machine has been given coded instructions which cause it to perform a task equivalent to that of the chart plotter and chart analyser before beginning the task of calculating the forecast. In some respects the difficulties of dealing with erroneous or mutilated teleprinter messages are greater than those of the mathematical calculations, but they must be overcome for, when forecasting by computer, there is no time to waste on the slow process of scrutinizing data by eye although for detecting errors and sorting data the human being is particularly well adapted. For example, one early forecast went sadly wrong because a ship had apparently reported its position as being in the middle of Greenland, probably an error in transmission. The human plotter, even the least intelligent, would reject a report of that kind, but to modify the machine's instructions in order to accept ships only on the sea is quite a little complication.

It is natural to ask how this new development is likely to affect weather-forecasting all over the world but the answer, raising matters of finance and national prestige as well as science, is not easy to give. It would, for example, seem foolish for many centres of the world to use computers costing perhaps £1 million to make the same calculations. On the other hand, if a country is not making or aiming to make the calculations, it is not likely to have the incentive for research and it may become totally dependent. Probably, matters will evolve in forecasting as they do in other activities with the smaller countries depending on the more powerful for what it is not economical to produce at home. As far as can be seen at present, the detailed problems of local forecasting of weather, of rain or snow, of fog or frost, as they vary over territories the size of, say, England, will never be economically solved at one centre, say in Washington or Moscow, responsible for disseminating the forecasts everywhere. This is because the central collection and later distribution of vast amounts of very detailed weather information would probably be more costly than the setting up of subsidiary forecast centres. We may then expect in a rational world some kind of cascade system with world-wide or nearly world-wide generalized forecasts, and long-range forecasts, being made at only a very few centres, with regional forecasts at several centres and with local forecasts produced wherever the interest is sufficiently demanding. The World Meteorological Organization is studying the technical aspects within its plan called World Weather Watch and one may be sure that the scientists will be ready when the statesmen have smoothed away the political and financial difficulties.

Perhaps the chapter should end on a realistic note. While forecasting by calculation has made progress it has still a very long way to go before we shall need no experienced forecasters for day-to-day work. As yet, numerical methods have done practically nothing for the southern hemisphere because the data from the upper air are too few over the enormous southern oceans. Also, they have done little for tropical regions for three reasons: lack of data, lack of scientific understanding of equatorial weather, and lack of effort, for research is expensive and manpower limited. And, finally, they have done little as yet to help with detailed local problems such as the visibility at an airfield or the thunderstorm at the garden party. It would seem that professional meteorologists able to make wise estimates from their experience will be needed for a good many years yet.

Chapter 11

The Winds of the World and the General Circulation of the Atmosphere

THE whole idea of climate arises of course from the fact, discovered early in the history of travel and exploration, that each part of the world has its characteristic régime of wind, rain, and temperature; a régime which varies from day to night, from day to day, from season to season, and from year to year, but nevertheless maintains by and large a steady character typical of the region. The task is then set to the climatologist to describe the climate of every part of the world in as much detail as may be interesting or useful, and to do so in a coherent and logical manner so that the knowledge may be communicated. A comprehensive reference work on climate will then present copious information in the two most effective media of communication, statistical summaries and geographical maps; for an overall view of the climate of the world the map has overwhelming advantages. To cover the variations through the seasons of the year it is convenient to prepare sets of twelve-monthly maps; not with the implication that the month as such has any particular relevance to climate but because some choice must be made and the conventional calendar month is for many purposes quite suitable, neither too long nor too short. One should perhaps remark that the original conception of the climate of a region as something quite definite and permanent, characteristic of the region, is no longer entirely tenable, and that it is now necessary to think of the climate itself as changeable. This behaviour, which will be the subject of a later chapter, is for some purposes important and then it may be necessary to speak not simply of the climate of a place, but of the climate in the twentieth century or whatever other period may be of interest. If, however, we limit attention to the last few hundreds of years and do not, for example, get involved with the last Ice Age, we may assume that the patterns of *average* world climate to be described in this chapter are in broad terms substantially permanent, although the variations of weather from one year to the next may be very considerable.

We begin with the average winds of the world and take upper-air illustrations adapted mainly from a comprehensive atlas prepared by the Meteorological Office. This gives world maps for the four mid-season months January, April, July, and October for a series of pressure levels in the upper atmosphere: 700 mb (near 10,000 feet or 3000 metres), 500 mb (near 18,000 feet or 5600 metres), 300 mb (near 30,000 feet or 9000 metres), 200 mb (near 39,000 feet or 12,000 metres), 150 mb (near 50,000 feet or 14,000 metres) and 100 mb (near 53,000 feet or 16,000 metres). It would be out of the question to reproduce or even summarize all the material of that atlas, but much of interest can be learnt from the few maps which have been selected for illustration. These are given in Plates 22–5 and apply to the two extreme months January and July, and are for 'sea-level' and the selected upper levels of 500, 300, and 100 mb. Rather than attempting to show the winds in speed and direction, the isobars only are given for the sea-level chart and the contour lines for the upper levels with a few arrows to indicate the wind currents. The general run of the winds is readily derived by recalling that the winds in the upper air flow along the contour lines with the lower values to the left in the northern hemisphere, to the right in the southern. It is clear that at all the heights illustrated in the upper-air maps the average wind in January is broadly from west to east in both hemispheres except quite near the equator, although with marked variations in direction in the northern hemisphere between north-west and south-west, and also marked variations in the closeness of the contour lines, and accordingly in the corresponding wind speed. In January at 300 mb (30,000 feet; 9000 metres), there is a ridge in the Pacific extending north to Alaska, a trough from north Canada south to the United States, another ridge in the east Atlantic stretching far into the polar seas, and a very pronounced trough again in eastern Asia extending south across Japan. Between the Atlantic ridge of high pressure and the east Asiatic trough there is a further trough-ridge pair of much smaller amplitude. In the southern hemisphere the features are far less pronounced.

Comparison between the January and July charts for the same level in the northern hemisphere brings out very clearly the relative weakness of the broadly westerly winds in the summer. In the southern hemisphere, by contrast, there is little difference between summer and winter. It is unfortunately impossible to represent the actual speed of the wind in small maps of this kind, but the cross-sections in Figs. 13 and 14 help a little to repair the deficiency. Fig. 13 gives a north–south section through the atmosphere in January at three different longitudes, at 10°W passing near the British Isles, at 75°E and at 140°E and so passing through the east Asian trough. The lines, so-called isotachs, represent in a very clear way the speed of the west–east wind through the section. Over east Asia the mean winds reach a speed of over 140 knots at the 200 mb level, the core of the average jet stream, whereas near the British Isles (60°N) the maximum is not much more than 40 knots and another maximum (the subtropical jet stream) is found at about 25°N, average speed over 60 knots.

The same longitudinal sections for the month of July show a radically different régime, the west-wind maximum over Japan having fallen to a modest 30 knots, not greatly different from that near the British Isles.

The variations in the strength of the average westerlies, with latitude, longitude, and season of the year, are perhaps the most striking features of the upper-air maps, but it is particularly illuminating also to fix one's attention upon the large region centred over northern India associated in everyone's mind with monsoons or seasonally different winds. North–south cross-sections for 75°E, that is through India, are also given in Fig. 13, 14 for January and July. In the winter time, northerly and north-easterly surface winds blowing outwards from the cold continent are quite shallow, and everywhere in the upper levels the winds are westerly. There is a pronounced jet-stream core with average speed more than 90 knots at a height of some 12,000 metres at about 25°N over India, that is definitely well south of the Himalayas. The change at all levels between January and July is quite radical. The surface pressure map for the summer months shows a low, the

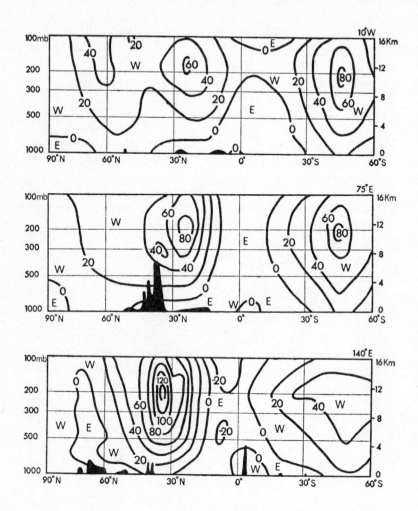

Fig. 13. January average cross sections from north to south showing strength of the westerly wind component in knots. Easterly winds are shown by negative values. Shaded areas indicate mountains. *Top,* 10° W; *centre,* 75° E; *bottom,* 140° E.

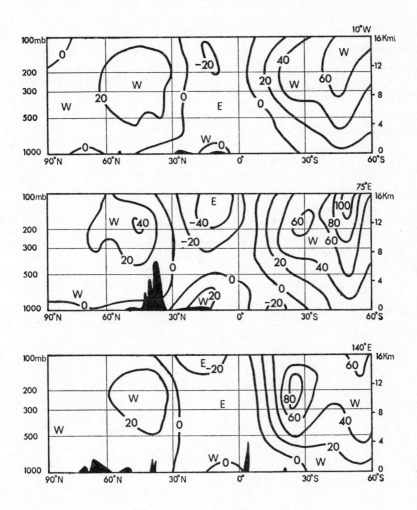

Fig. 14. July average cross sections from north to south showing strength of westerly wind components in knots. Easterly winds are shown by negative values. Shaded areas indicate mountains. *Top,* 10° W; *centre,* 75° E; *bottom,* 140° E.

so-called monsoon low, centred over north-west India with winds generally south-westerly over the peninsula. These are generally deeper than the winter north-easterlies but not very deep: at a modest height over India, 4000 metres more or less, the winds actually change to easterly and increase steadily with height to give an easterly jet-stream core not far north of the Equator. The westerly jet in the northern hemisphere is much weaker than in winter and is displaced to the northern side of the Himalayan mountains. Looking at the upper-air map for 100 mb, about 17 kilometres (Plate 25) we see that the contour lines define a 'high' centred almost precisely in the same geographical location as that where the 'low' is found on the surface map in Plate 22; the upper winds, easterly to the south and westerly to the north of this high, are part of a major circulation system covering a large area of Africa and southern Asia. Variations of a similar kind from winter to summer, although on a smaller scale, are to be found also over tropical North America, and in the southern hemisphere, over tropical South America and South Africa. The general characteristic is a summer high pressure centre in the higher levels of the troposphere vertically above a region of relatively low pressure near the ground with, of course, an anticyclonic wind circulation overlying the generally cyclonic mode at the surface. Putting it in a rather different way, one may say that in the upper troposphere and lower stratosphere monsoonal (that is seasonal) wind changes between winter and summer in tropical regions, tend to be in a direction opposite to the changes near the ground – a dynamical structure of considerable interest.

The general circulation of the atmosphere

It was from the study of the prevailing winds as they were observed, especially the surface winds from the days of world exploration in sailing ships, that the concept of the general circulation of the atmosphere, the most important concept in the whole of meteorology, arose. The obvious question presenting itself to the inquiring mind is to explain why the winds are as they are, where the trades, the brave west winds, the monsoons, and so on come from, where they go to and how the whole system of low-level and high-level winds is connected into a perpetual three-dimensional circulation.

This is essentially what is meant by the problem of the general circulation, a matter for speculation over the last 300 years.

We shall, however, resist the temptation to digress into the pleasant byways of history, for the early ideas on the theory of the general circulation, even through the nineteenth century, were little more than scientific curiosities. The twentieth century has been a period when the remarkable facts of upper-air temperatures and upper winds have accumulated too quickly for the few scientists working in theoretical meteorology to keep abreast, and only in very recent years has the gap begun to close. And even today there is no well-rounded theory to explain why it is that the pattern of the general circulation is as it is observed to be, and not something quite different. But we do have a fairly complete descriptive account of the system, with a good deal of theoretical knowledge on how and why the system works, and this we shall briefly examine.

If the general circulation is, by definition, a matter of the average winds of the world it immediately embraces all the other major physical quantities and physical processes. One cannot go far in the study of winds without introducing the driving force, the pressure; and one cannot go far in the study of pressure without introducing air density and air temperature for the pressure is the weight of the air above. Again the study of temperature introduces energy and its transformation, the absorption and emission of radiation, evaporation and condensation and the conversion of latent heat: vertical motions introduce clouds and rainfall, and so on until the whole of weather science is drawn into the argument. Thus, a complete understanding of the general circulation of the atmosphere implies a complete understanding of every process in world climatology and it is for this reason that it may justifiably be called the central problem of the science.

To describe and explain the average pattern of winds all through the atmosphere is rather like preparing an annual balance sheet for a complicated business in that it is implicit that every transaction has been taken care of; and the analogy is not entirely fanciful for in analysing the general circulation it is necessary to establish that certain accounts do in fact balance. The laws of physics are essentially conservation laws: matter must not disappear which means that over any region the inflow and outflow of air must in the long run balance; water must not disappear without reason so that evaporation must be sensibly balanced by precipitation; energy is conserved and if the earth and its atmosphere are in much the same state in one year as in another – and the climatic changes are certainly very slow – the energy gained must be balanced by the energy lost. The motion also is much the same – the general circulation is a permanent and rather steady organization – so that the losses of momentum of the wind due to friction with the earth must somehow be made good. A full theory of the general circulation must then begin with the fundamental laws of physics and derive from them, implicitly or explicitly, complete geographical balance sheets for the annual working of the atmosphere in all its different currencies – a tremendous undertaking which has not yet been accomplished.

Viewing the matter from a rather different angle we may say that the earth's atmosphere has a complex pattern of structure and behaviour – in terms of winds, pressures, temperatures, humidities, etc. – which is characteristic of the earth and so far as we know unique in the universe. The whole may be called a thermodynamic physical 'system', it may be called an 'organization' and, with no great stretch of imagination, an 'organism', suggesting an analogy with a biological organism. It is an analogy worth pursuing a little way because the methods of investigating the climatological organism have had something in common with the methods of biological study and, so far, rather little similarity with the mathematical deductive methods of 'solving' problems in physical science.

In coming to grips with the puzzles presented by the behaviour of a biological organism, a plant, or an animal, one does not, in the present state of knowledge, expect to sit down with pencil and paper and from a set of basic laws and assumptions to conjure up the complete organism as a logical necessity. Rather one takes the organism as it is found, observes its structure and behaviour and tries to describe, and in some terms to explain, the mechanism. Numerous general problems are formulated: how is material substance circulated through the living organism, how is energy supplied, distributed, and consumed, how is the temperature regulated? Numerous, more specific, functional problems arise: what is the rôle in the general economy of the organism of this anatomical feature, of this gland, or of that organ? By this piece-meal approach a general understanding of the whole is gradually achieved – and so it has been with the general circulation of the atmosphere. How is the momentum regulated; how is the energy supplied, distributed, and disposed of; where does the moisture originate, what processes convey it over the world to be deposited elsewhere as rainfall; how is the curious distribution of temperature in the atmosphere, with its troposphere, stratosphere, mesosphere, and thermosphere, to be understood? And among the specific problems of function, one may ask what is the importance to the general economy of the thousands of thunderstorms occurring every day – do they affect the general circulation? Do the passing depressions and anticyclones play a basic rôle or are they just so much embroidery on the pattern? Is the distribution of mountains, and of land and sea generally, of controlling significance or would the climate be fundamentally the same on an earth of uniform surface? Do the ocean currents significantly affect the climate or is the benefit derived from a warm Gulf Stream bathing the coasts of western Europe only a myth? To go further afield, could the ozone in the upper stratosphere modify the general circulation and the climate of the world or is it just an interesting curiosity in a complex system?

In attempting to provide convincing answers to questions of this kind a wealth of detailed knowledge on the general circulation of the atmosphere has been accumulated, and a few items are selected by way of illustration.

The water balance The earth's atmosphere at any time contains an amount of water vapour which, were it all condensed and deposited on the surface of the earth, would stand to an average depth of 2·5 centimetres, an inch. Its distribution is far from uniform, the averages for July and January being as represented in Plate 26: most of the moisture is in low latitudes with India as the area of summer maximum. Over the earth as a whole, something like 100 centimetres of rain falls each year. The amount is not accurately known, not better than within 10 per cent, because no good method of measuring rainfall over the open sea has yet been devised; (it is rather surprisingly easier to measure how much water vapour there is in the atmosphere above than to measure how much falls out on to the surface below). However, for our present purpose 100 centimetres is near enough. The moisture in the air at any one time therefore represents about one-fortieth part of the annual rainfall, say ten days' supply. The average available vapour does not correlate very closely with the rainfall. For example, in some desert regions the air contains considerable water vapour but it never precipitates, whereas elsewhere, especially over mountainous areas in middle latitudes, the air may contain on average no more than a day's supply of rain. These geographical differences are balanced out by air movements, by the winds, but by and large the moisture in the air needs to be completely replenished by evaporation every ten days. It is a sobering thought that all our rainfall must be drawn from the atmosphere which works with only a ten-day supply in store: but fortunately the replenishment by evaporation over the earth as a whole is nearly as steady and reliable as the sun which provides the heat for the process.

The mechanism of the global water cycle has many interesting features, a number of which will be revealed if we take a quick run through the cycle. Where one begins is not material but perhaps evaporation is as good a starting point as any. Evaporation takes place more or less continuously from all parts of the earth's surface but at very varying rates. There are three esssential factors in the process: first, a supply of water at the surface; second, air which is not already saturated; third, sufficient heat energy to support the evaporation, that is to supply the latent heat. Naturally the importance of the different factors varies according to circumstances and there are particularly radical differences between land and sea. A land surface may be quite wet with moisture freely available for evaporation; or it may be quite dry like the desert or, indeed, like parched land anywhere after drought; or it may be somewhere in between as is the normal condition when much or all the moisture has to be evaporated from vegetation, drawing its supply through the roots. The available heat energy is the main limitation and because soil and ground do not store heat efficiently the energy must be supplied from sunshine at or near the time of evaporation. For this reason evaporation over the land is a maximum in the summer time and may become almost negligibly small in the winter in middle and higher latitudes. It is also a day-time process and often ceases completely in the evening to be followed by the reverse process, the deposit of dew, during the night. All this may seem a far cry from the general circulation of the world's atmosphere but the detail serves to point the moral, that one cannot explain the broad features of world climate if one does not know the actual mechanisms involved. Evaporation from the land, taking the year as a whole, is often interrupted and limited, sometimes by lack of water, sometimes by lack of heat.

Over the sea the circumstances are quite different. In the first place there is never any lack of water. In the second place, sunshine – or solar heat – penetrates the ocean in depth and is stored so that in providing heat for evaporation the input at the time is not important. The critical factor here is the dryness of the overlying air, specifically its dewpoint relative to the temperature of the sea surface, and evaporation takes place most freely when cold dry air blows over a warm sea for then convection currents will carry the evaporated water upwards and the process will be maintained. Whether it is day or night is of little moment, while the season of most intense evaporation is the winter when the heat stored during the summer is drawn upon most rapidly to warm and to moisten the cold winds. Some parts of the ocean are particularly important as suppliers of moisture to the atmosphere as is revealed in the map of Plate 16 which, it may be noted, is entirely an inference from theory for there is no way of directly measuring the evaporation which actually takes place from the sea. Notable centres of high evaporation are off the east coasts of the main continents where cold, dry winds commonly blow in winter from land to sea, while the extra heat in the sea required to main-tain the high rate of evaporation is in part transported to the area in the warm ocean currents – the Gulf Stream off the American coast.

Evaporated water must find its way back in due time but the geographical pattern of precipitation has little in common with that of evaporation and the next stage after evaporation is transport in the air, transport by the winds to the rainy locality. Thus, the heavy rain of the tropics is largely provided from moisture evaporated from the sea into the trade winds or the monsoon winds which may have themselves been directly derived from the trades. The north-easterly and south-easterly trades blowing from higher to lower latitudes are relatively cool and dry for their latitudes and are avid absorbers of water from below. The rains of the western continents in middle latitudes feed upon the moisture evaporated from the oceans to the west, a relatively reliable source in the belt of prevailing westerly winds. The eastern parts of these continents in similar latitudes are more dependent on moisture in winds from the east, climatologically unreliable, and implying a consequently greater liability to disastrous droughts.

The next stage is the rain-forming mechanism itself, the processes by which the water vapour is condensed into clouds and combined into drops or flakes large enough to fall to the ground, and these we have discussed in earlier chapters. The main mechanisms are vertical convective systems, that is showers and thunder storms, and cyclonic storms and depressions, much enhanced in their efficacy as rain producers by favourable topography and especially by notable mountain ranges. The pattern of rainfall over the world depends therefore only in part on the occurrence of moisture-bearing winds for there is no virtue in moist air if there is no mechanism to extract the moisture. The dynamical rain-producers, found preponderantly in some geographical areas and hardly at all in others, give us a world climate with its rainy and dry regions which everyone should have learnt about in the early years of school geography.

This is an appropriate place in which to emphasize one of the most significant facts about the general circulation. The water cycle does not work without various cogs in the transmission, and one vital cog is formed from cyclonic depressions and convective showers. It is not reasonable to attempt an explanation of the general circulation on the assumption that the rain-producing mechanisms, small as they are compared with the earth's atmosphere as a whole, are a non-essential feature, a kind of modern convenience for the comfort of mankind. One might as reasonably expect to understand the running of a motor-car on the supposition that the purpose of the engine is to supply warm air for the heater.

The earth's water cycle has many further features as, for example, the storage underground, so important in water conservation, and, in much larger quantities, the storage in the great ice-fields of the Antarctic continent and Greenland. It has been estimated that the water locked up as ice would, if melted, be sufficient to increase the depth of all the oceans by 100 metres and, knowing that all the ice has disappeared for long periods in the long history of the earth, we may suppose that it may well do so again. But the time scale of significant variation is that of climatic change, so that constancy in ice storage is a reasonable assumption for our present purpose.

In any case, the amount of storage of water in the atmosphere is so small that over a period of a few years almost perfect balance between evaporation and precipitation is attained, taking the earth as a whole. Some regions, as we have seen, have an excess of evaporation whilst others an excess of rain, and the unevenness may be looked at from two points of view. First we may think of the balance within the atmosphere and note again that geographical excesses and deficits are made good by transport in the winds. Thus, an area with excess evaporation, for example the western temperate oceans, will export moisture in the winds while one of excess rainfall, such as the western temperate continents, must import moisture in the winds.

Equally valid, however, and of more obvious human interest is the view point of the water-user concerned with the water balance of the ground and to him an excess of rainfall over evaporation is mainly balanced by run-off which goes to feed the rivers of the world, whereas an excess of evaporation over rainfall can in the long run be made good only by irrigation. Generally speaking, that is to say with the exception of limited regions such as the Nile valley which depend for their water upon irrigation from remote sources, rainfall over the land is on average over the world about 30 per cent in excess of the evaporation; the balance feeds the rivers which top-up the oceans once more. The details of water circulation over the surface of the land and underground falls to the important science of hydrology, and the water balance within the oceans to oceanography, both beyond our present scope, but a brief digression as far as the Mediterranean may be permitted. The rivers flowing into this inland sea do not suffice to make good the evaporation and water is continuously drawn from the ocean through the only channel, the Straits of Gibraltar. This is of course salt water whereas evaporated water is pure and, consequently, a continuous increase in salinity would be expected. It appears, however, that in the long course of the ages a balance has been reached with Mediterranean water now more salt and more dense than Atlantic water but no longer becoming more so. This state of affairs is possible only because some of the heavy Mediterranean water flows out at lower levels, giving in the state of near-balance a net inflow of water but no net inflow of salt. Interesting self-regulating mechanisms of this kind abound in natural science as indeed they must, since what persists evidently can persist, but they introduce difficulties into very many problems where a slight departure from balance, a slow trend one way or another may be the focus of interest. In a complicated problem of balance it may be very difficult indeed to make measurements – as of rainfall, evaporation, run-off and the like – sufficiently accurately to verify what we already knew was true, that, so to speak, two and two make four. To go further and show that there is a very slight remainder unaccounted for may seem almost impossible – and so far, it often is.

The balance of energy in the general circulation of the atmosphere presents a problem of this kind, to which some attention was given in an earlier chapter. Here we need only remind ourselves that to maintain the earth's atmosphere in much the same state over the years the energy input and output must balance not only globally, for the earth as a whole, but locally too, for every region separately. The complete balance sheet is formidably complicated and has never been attempted in geographical detail, although a number of the components have received laborious attention and resulted in maps, for example, of incoming solar energy, radiation exchange at the earth's surface, and latent heat input to the air which is practically identical with the map of world rainfall.

Much attention has also been given by meteorologists to the balance of momentum for the same general principle applies, and it must be true that over any region the processes which have caused the air movement to change must, in the long run, cancel each other out. The interesting question is to identify the significant processes and most attention has been devoted to the momentum in the direction parallel with the lines of latitude or, a more convenient but closely related quantity, the angular momentum of the winds around the earth's polar axis. As we have seen, the winds of the world blow more or less in this manner round the axis with varying speeds and keep going year after year: by what process?

One may present a more specific problem by imagining a ring of air surrounding the earth and bounded by two chosen parallels of latitude, say 40° and 45°N. Although the wind does not blow precisely in the west to east direction and varies with time, there is always a definite value to the total angular velocity of the ring. The only external force is gravity which has no effect in that direction. There is also the fluid pressure, but this is an internal force with no effect taken round the closed ring. At the earth's surface, in this chosen latitude, the winds in the ring are mostly from a westerly direction and the frictional drag should cause the spin in the ring to decrease. The only way in which the momentum can be kept up against this frictional drag is by gaining momentum in the exchange of air which is always going on between the air in the ring and the remainder of the atmosphere to the north and to the south. From world maps, such as are illustrated in Plates 17–21, it is possible to compute how much momentum passes across the boundaries at the chosen latitudes according to these average winds and it is found that it is insufficient to balance ground friction. This difficulty can be overcome only by doing the arithmetic in much more detail, not on the basis of average winds, but on the basis of the actual winds day by day. The conclusion is not trivial, but proves that it is impossible to explain the winds of the general circulation by studying only the average winds of the general circulation itself: it is essential to take account of the daily fluctuations, in effect of the varying winds of passing depressions and anticyclones with their troughs and ridges in the upper atmosphere. This fact was not clearly understood until the nineteen-twenties, and all 'explanations' of the general circulation before that time were necessarily unsatisfactory. We should add that there is no special virtue in the ring at 40–45°N chosen by way of illustration, and the same considerations apply to any zonal ring in either hemisphere. If, however, a ring is chosen near the equator in the belt of the easterly trade winds, surface drag evidently tends to a steady loss of spin from east to west, or a gain in the spin from west to east, which must be made good by exchange with neighbouring rings. When, however, the whole atmosphere is taken together there is no way of gaining momentum by exchange, and the only possibility is that the surface drag just cancels out in the long run. In other words, whatever the general circulation may be, one control must be satisfied: the easterly and westerly surface winds must balance in the long run in such a way that the total friction with the earth exerts, on the average, no force around the earth's axis. It may seem odd that the prevailing westerly winds of the Atlantic between Europe and America are possible only because there are easterly winds in the trades farther south, but some such conclusion is inescapable.

The general circulation of the atmosphere, it now turns out, does not consist of easterlies in one part and westerlies in another, each self-contained, but is one overall system, one interlocking mechanism which includes the whole atmosphere. Naturally, it is complicated by the land and sea distribution, by mountain barriers and by seasonal changes, and in order to clarify the problem for theoretical study attempts have been made by many experts to reconstruct what they imagine the general circulation would be like in the absence of such complications, that is to say on a symmetrical earth with its axis of rotation perpendicular to the plane of the ecliptic (to eliminate the seasons) but otherwise similar in physical conditions to our real earth. The idealized general circulation model for a symmetrical earth has been inferred by taking the observed facts from the real earth and, largely by intuition, discarding what seems to be geographical or seasonal complication and retaining only what seems to be fundamental to the planet. In this effort of abstraction, great weight is given to the evidence from the southern hemisphere which, in the absence of large continents, may be expected to display the features of the uniform earth relatively clearly. Also we have the evidence from the equinoctial months or we may take averages throughout the year and in this way minimize the seasonal differences. On a symmetrical earth there would of course be no basis for distinguishing one longitude for another, and the climate or general circulation would be uniform around each latitudinal belt. The divisions of climate would then be into zones from the equatorial belt to the polar caps, and the number of separate zones which it would be profitable to distinguish in a classification of climates would be a matter of convenience. In Fig. 15 we have illustrated one version of this idealized general circulation in order to bring out the most significant features.

Nowhere on the planet must one look for a permanently steady

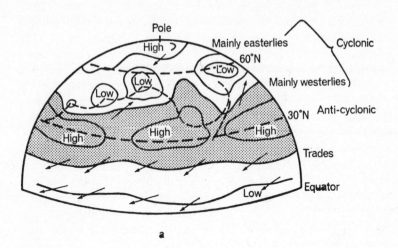

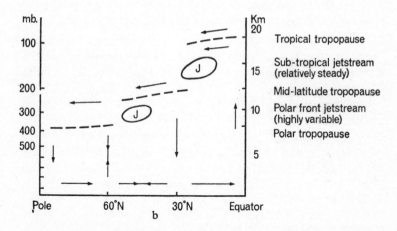

Fig. 15. Schematic representation of the general circulation of
the atmosphere on an idealized earth, ignoring the effects
of continents and oceans and the seasons. Upper diagram
shows the surface pressure pattern as it may be on one
day. The long period average would show simple zones
of low and high pressure. The lower diagram represents
a vertical cross section from pole to equator.

145

pattern of behaviour. Everywhere we must have variability, changing pressure, veering and backing winds, and so on, but some zones we infer would be much more disturbed than others. The greatest variability lies in middle latitudes, say from 40 to 70°N and S, affected by a succession of cyclones (depressions) and anticyclones or ridges. The cyclones are baroclinic, or frontal disturbances developing in the zone of strongest temperature gradient between the cold air of the polar cap and the warm air of the tropics. As a consequence of their dynamical development and movement, the cyclones move first towards the east and then as they deepen trend polewards attaining their maximum depth in the belt, say between 60 and 70° where they gradually fill up. The anticyclones or anticyclonic ridges which develop between successive depressions in the same middle-latitude baroclinic zone, also move from west to east but, as they develop, tend to move out of the zone into lower latitudes maintaining a permanent belt of high pressure cells in subtropical latitudes.

We must think then of these unsettled mid-latitude zones as providing the main engines to drive the whole circulation. The region of strong thermal gradient with strong westerly winds in the upper air, culminating in the polar-front jet stream near the base of the stratosphere, is permanently unstable giving birth to the succession of eddies near the surface, throwing the cyclonic eddies poleward, the anticyclonic eddies equatorward and so maintaining the belt of disturbed surface westerlies. Over the polar cap, beyond the graveyard of the depressions, the surface winds are drawn from the east whereas on the equatorial side of the subtropical anticyclonic belts, which may, if we care for the metaphor, be called the graveyards of the baroclinic anticyclones, easterly surface winds must also develop to converge from the two hemispheres towards the equator.

The mechanism of the equatorial cell needs next to be described for it forms in effect another engine driving the circulation, or two engines if we think of the hemispheres separately. The air converging in the trade winds from the north and south is heated and moistened continuously at the bottom and eventually becomes unstable. The instability is, however, not of the same kind as that within the baroclinic westerlies which are unstable because warm and cold air masses lie side by side. In the equatorial regions, the instability is due to the existence of warm moist air in the lower levels below cooler upper air; there is vertical instability, and the typical outcome is the heavy convective shower or thunderstorm penetrating throughout the troposphere to the base of the stratosphere. This form of upward thrust, driven mainly by the latent heat liberated in the formation of rain, conveys large parcels of air from near the surface and deposits them, so to speak, at the base of the stratosphere from which region they can escape only by flowing polewards at this higher level. The thunderstorms and showers, of which there are thousands in action at all times, act rather like a series of large explosions each forcing a portion of surface air to the higher level and, together, giving a net upward current as indicated in Fig. 15 (b).

The two engines in each hemisphere of equatorial and middle latitudes are each interesting in themselves, but perhaps the most subtle feature of all is the way in which the units are geared together to form one world machine. A great deal of research has gone into this aspect of atmospheric mechanics and still there are areas of uncertainty and argument, but a very important link seems to lie in the upper troposphere in the latitudes of the subtropical high-pressure belt. Looking at the diagram we see that this region is marked as carrying the subtropical jet stream. It is fed from low latitudes by the high-level outflow from the equatorial convection, and the strong westerly current of this jet stream is accounted for by the effect of the earth's rotation acting on the poleward moving air. In these subtropical latitudes the current forms the boundary to the unsettled baroclinic westerlies, and is continually disturbed by the wave-like distortions of these westerlies with the result that the upper winds of middle latitudes continually extract westerly momentum from the sub-tropical jet, at the same time as portions of warm tropical air at lower levels are captured and carried away to form the warm sector of another baroclinic depression. Middle-latitude westerlies feed upon the westerly momentum created in low latitudes and upon the energy of the warm air in the subtropics. The baroclinic disturbances bring westerly momentum to the surface of the earth where it is frittered away by friction. The outbreaks of polar air which must balance the sectors of warm air pass into the subtropical highs, and from these the trade winds are drawn. Moving from high to low latitudes the trade winds would, in the absence of surface friction, develop momentum from the east and, in moving from – say – 30°N or S to the equator, winds would increase to over 100 miles per hour. Friction, however, dissipates all this easterly momentum and air arrives at the equator with little motion and is able after rising to high levels to move polewards. The poleward movement, by the conservation of angular momentum, implies increasing westerly winds which maintain the subtropical jet stream and the supply to the mid-latitude westerlies to complete the circle. The diagrams of Fig. 15 should help to fix ideas on this model of the mechanics of the

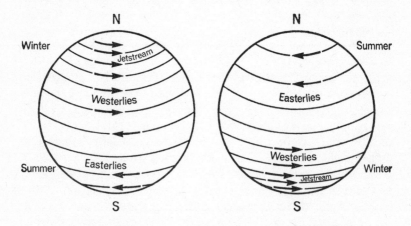

Fig. 16. Monsoons in the stratosphere. At high levels, 20 to 35 miles above the earth, there is a complete change from season to season with westerly winds in the winter hemisphere and easterly winds in the summer hemisphere. Near the edge of the polar winter darkness a stratospheric jet stream develops and disappears suddenly in the spring, with sudden warming of the stratosphere.

general circulation which, hidden in the intangible variability of fluid motion, has been difficult to establish from the observations, but is not particularly difficult to grasp after a little familiarity.

If a mechanical engineer were to be introduced to a complicated and unfamiliar working engine, or an electronic engineer to some elaborate circuitry, he might set about the task of understanding the system by studying the function of each part, much in the way that the meteorologist has studied the atmosphere, and satisfying himself that his diagnosis led to a consistent model; this done, the engineer could fairly say he had solved his problem. To the meteorologist, however, a knowledge of how the general circulation works, important and useful though this is, is not an end to the matter. There remains the much deeper question as to why the atmosphere surrounding the earth should as a result of heating by the sun and cooling to space adopt this particular pattern, this highly specific mode of circulation, and not any other among the infinity of conceivable patterns. Some progress has been made towards the solution but most of the work remains for future research to accomplish and we may conveniently leave the matter here.

Chapter 12

Long-range Forecasting

To anyone, scientist or layman, who had not studied the subject it might appear reasonable to suppose that techniques of weather-forecasting could in principle be extended indefinitely into the future, and that the period of time for which useful forecasts could be made would be increased gradually with the advance of scientific knowledge. The facts do not, however, bear out the supposition. Short-range forecasts for about twenty-four hours ahead have been made throughout the century of forecasting and today the usual range is much the same. It is true that 'further outlooks' or 'medium-range' forecasts indicating the general character of the weather are issued in some countries for a few days ahead, but in those parts of the world which are subject to weather that is changeable from day to day it is not worth while to attempt precision in the timing of the changes beyond twenty-four or thirty-six hours.

There is, it seems, a natural limitation to the period of validity of detailed forecasts, a kind of time-barrier through which it is proving extraordinarily difficult to penetrate however much we may advance in scientific understanding and it is interesting to note why this is so. Detailed weather-forecasting implies and depends upon the prediction of the pattern of winds and weather, as may be represented on geographical maps. Predictability depends on the structure of these patterns and upon the degree to which the significant features may be traced coherently as they form, develop, move, and decay. Now it so happens that the most important features which control the wind and weather over most of the world, are the travelling depressions and anticyclones, systems which develop as entities within a matter of hours and go through a typical cycle of growth, movement, and dissolution within a few days. If one takes any weather-map of, say, the Atlantic area, there will be one or more cyclonic centres each of which is readily traced on earlier or later maps, but which is lost in the general pattern within a day or two, in the past and in the future. Forecasting depends upon the skill in predicting the positions, intensities, and structures of all the relevant systems and it is one thing to predict the behaviour of the systems which exist, quite another to do so for those which do not. It is hardly more reasonable to predict the future of an unborn depression than to predict the career of an unborn baby and, by the same token, it is not possible to make detailed forecasts of weather if there is the likelihood of events being affected by a new development – and in most parts of the world for most of the time the limit is about twenty-four or thirty-six hours. Sometimes of course there is no threat of any new disturbance, particularly if one's region is within the area covered by a large established anticyclone, in which case one may venture to forecast little change for a few days. This was true in the early days of forecasting when, although the theory of depressions and anticyclones was not at all well understood, their movements could be extrapolated by pure empiricism: it is still true today when the theory is known and the changes may be scientifically calculated – the natural time-barrier remains where it was. According to present scientific thinking there seems little prospect of ever breaking through this barrier so that no meteorologist of standing will encourage the public to expect weather-forecasting even for two or three days ahead to be made consistently accurate with the detail to which they are now accustomed for one day ahead. Research holds the promise of many things, but not of everything, and long-range forecasts of weather changes day by day is not at present on the agenda.

It is, however, possible by essentially the same methods, that is by estimating the changes in the geographical patterns of wind and weather, to say something useful about the future for some days ahead provided the detailed treatment of individual depressions and rain areas is not expected. There is a very large-scale pattern determined by the large anticyclones and major cyclones on the surface weather-maps and by the 'long waves', the major sinuosities, in the circumpolar flow of predominantly westerly winds. This large-scale pattern changes relatively slowly and determines the trend in the general type, whether settled or unsettled, generally warm or cold. There is therefore, a scientific basis for 'further outlooks' or 'medium-range forecasts' for a few days ahead, but thereafter the evolution of the pattern even on the broadest scale cannot yet be predicted with useful precision and this method of forecasting finally breaks down.

It is at this stage of failure in the techniques of short-range forecasting that long-range forecasting takes over, using a variety of methods which purport to give some clue, better at least than sheer guesswork, to the nature of the weather in the weeks or even months ahead. Very useful information about seasonal changes is of course provided from pure climatology, and in one sense we make a useful forecast every time we prepare for the warmth of summer or the cold of winter, using as a guide to the future only the experience of the weather at the same season in previous years. But in its usual sense a forecast must be something more than climatology; it must be something specific for the unique occasion, and not a statement which would be equally valid for one year as for another. A long-range forecast must then at least indicate how the weather over the extended period will differ from the climatic normal. It may of course do more than this but many, indeed most, systems of long-range forecasting do in fact attempt nothing more than an indication of the degree to which the mean temperature and rainfall will differ from the long-term average, and the poor standard of accuracy often attained when aiming at this modest goal may not encourage more ambitious schemes; but it is not the only approach. Climate does not consist only of long-term averages, and the idea of climatology as the listing of average values of climatic elements for all parts of the world was never accepted by climatologists as more than a beginning. The climate of a region and season is a summary of all the types of weather conditions which occur, and a long-range forecast may attempt to describe the future weather in terms of the types of synoptic weather-situations to be expected.

That there is something which it would be interesting – and economically useful – to be able to predict, needs no technical knowledge to demonstrate. In spite of the apparently wide variations in the degree to which the economic life of a country depends upon its weather conditions, long-range forecasting could be helpful almost everywhere. The human race has always needed to adapt its way of life to the climate and, as is the way of such things, adaptation has usually meant fitting in well with the normal occurrence and putting up with the unfavourable extremes. Thus, in all parts of the world, whatever the climate and whatever the way of life, extreme climatic conditions are generally un-favourable conditions, the worst consequences of which could be avoided or at least minimized if they could be predicted with adequate notice. The more advanced the civilization and the more independent of weather the life of the citizen may appear to be, the more would be the value of long-range weather-forecasts for more could be done by planning ahead. It is only necessary to think of one or two examples to drive the point home and perhaps power supplies, transport, and food are suitably important. The power requirements of a highly-developed country are dependent upon the weather conditions to such an extent, that a severe winter may strain resources to the point of breakdown. At all times good management calls for no more than the necessary and sufficient insurance against excess demand, for an unnecessary margin of safety is waste. Whether explicitly or not, a weather element is built into all forward planning, on stocks of fuel to be built up or held in reserve, on the phasing of the development of new plant, on maintenance schedules, and on the margin of stand-by plant to be available for contingencies. Without further elaboration it is obvious that a reliable long-range forecast for weeks, or better, a season ahead could affect the planning of power, transport, and food production in such a way as to be a significant item in the budget. Whether the item would represent a few per cent or only a fraction of 1 per cent in costs, the value to the community of good forecasts wisely used could run quickly into many millions of pounds annually. The cost of producing long-range forecasts, if they could be reasonably reliable, would beyond any doubt be entirely trivial compared with the saving made possible by greater efficiency, by the reduction of waste, in agriculture, industry, and commerce. The man in the street no doubt thinks of weather first in terms of holidays and recreation and, for the age of affluence and leisure which we are so often assured lies not far in the future, these less serious matters are not to be brushed aside as frivolous; but, at the present time, good long-range forecasting is not a mere entertainment or amenity in a welfare state, it is an aid to business efficiency worth a great deal of hard cash if it can be obtained.

It should, however, be admitted at once that up to the present time long-range forecasting anywhere in the world has been no more than marginally successful in so far as there is published evidence to go upon. Various statistical and other methods have been used in different countries from time to time. In nearly every case the work has been the interest of one enthusiast, working alone or with a very small team of assistants. The specialist has convinced himself that his techniques are sound and his forecasts reasonably successful (how else could he have continued), but to convince other scientists that real progress has been made has proved more difficult. Curiously and significantly, it is hard to find in the history of this tantalizing subject an example of a technique of long-range forecasting which has been developed in one country or one institution and has been taken over as a working system anywhere else. This of course is quite out of line with the usual way of scientific progress in which every step forward is liable to be exploited elsewhere, with almost indecent haste, as soon as it is known; it is, in particular, quite out of line with progress in short-range weather-forecasting which all over the world has followed a uniform pattern, almost to a fault.

The explanation of this spasmodic and often abortive activity in long-range forecasting is of course to be found mainly in the small degree of practical success which has attended the methods. Authorities have been discouraged from investing substantial resources in an enterprise which carries the risk of throwing good money after bad. But this is not the only factor. For the research scientist, especially for the independent thinker in the universities, the obviously complicated and intractable nature of the problem, regarded from the theoretical view-point, coupled with the prodigious problem of data handling, has been repellent. Meteorological groups in universities have been small and the prospect of devoting much effort over many years to a project which may bring credit to no one is never attractive.

The theoretical difficulties are indeed formidable and up to the present time no one has tried to explain, beyond the level of speculation, why it is that the weather conditions in one place and season may vary so much from one year to the next. Why in 1959 did western Europe experience one of the finest summers of the century, in 1960 have one of the worst, and in 1962–3 break all records for persistent winter cold since observations began? Quite frankly we have hardly the glimmerings of a theory, nothing more than conjecture, a listing of possibilities, and until the facts are much better understood than they are at present, long-range forecasting will remain almost blindly empirical depending upon the discovery of working rules by any methods available.

No great interest is to be found in a record of the largely unsuccessful gropings for prognostic relationships, but a brief note is needed if only to show that many possibilities really have been examined, especially by statistical methods. The study of periodic variations in the weather has attracted many people and some major efforts have been made, but so far there is no period firmly established other than the two obvious ones, the day and the year. Only very recently workers in the United States and in Australia have claimed to discover a genuine relationship between the rainfall and the phase of the moon, and if the claim stands up to further critical examination many meteorologists will need to eat their words and offer apologies to the sailors and the shepherds – but not even the most ardent of the new moon worshippers would expect the relationship to give much useful help in forecasting. The sunspot cycle, which averages something over eleven years, is another favourite, and not without reason for there is quite certain knowledge that the high atmosphere, including the ionosphere, is much more disturbed in periods of high solar activity than in 'years of the quiet sun'. The last phrase is topical in that the years 1964–5 have been selected by geophysicists for special international study as the IQSY, supplementing in an important way the work of the famous International Geophysical Year, 1957–8, when solar activity was at a very high level, the highest since records began. But, however important the variations may be for the very high atmosphere, the significance for weather near the ground is not yet established. Some workers have satisfied themselves that a connection exists and Professor Baur of Germany, who has the credit of having practised long-range forecasting perhaps longer than any other scientist in the world, is a convinced believer in the importance of the phase within the sunspot cycle. But in slipping into the use of the word 'believer' one concedes that faith is as strong as scientific proof in this context.

If true periodic variations in weather do exist, it is probable that they contribute too little to the overall variations to be of prediction value, but this conclusion does not mean that there is no rhythmical behaviour whatever. For stretches of time, weather conditions over a large area may get into a rhythm and vary from, say, wet to dry or generally cold to warm conditions within a few weeks; if the rhythm can be recognized it may sometimes be trusted to continue long enough to give guidance in forecasting.

The use of standard correlation methods for weather prediction has also been used and, in the case of the attempts to forecast the rainfall of a monsoon season in India, apparently with some success. Sir Gilbert Walker, who was the head of the Indian meteorological service from 1904–24 and a very able mathematician, was responsible for this early work; but more recently, one learns that the methods are failing to maintain their early promise. The application of correlations to prediction rests on the success in finding associations between one meteorological factor, temperature or rainfall as a rule, and the same or some other factor at an earlier time, perhaps six months earlier if seasonal forecasting is in mind. Walker discovered, for example, an apparently significant relationship between the accumulation of winter snow in the Himalayas and the strength of the following summer monsoon, a relationship for which a plausible physical theory might readily be provided. But many other relationships, all of them rather weak, were combined together to provide a prediction of the seasonal rainfall. It is recognized today that searching for correlations, without good physical reasons to guide one, is a hazardous undertaking easily leading to 'discoveries' which later turn out to be accidental coincidences of no prediction value. So far as is known, correlation methods are not systematically used as a method of long-range forecasting anywhere in the world at the present time.

To close this chapter, a description will be given of the methods of long-range forecasting for a month ahead as developed in the Weather Bureau of the United States and in the Meteorological Office of the United Kingdom, these being by far the most ambitious attempts that have yet been made. Pride of place must be given to the work in America which has provided monthly predictions twice monthly since 1948.

The main feature of the Weather Bureau's system, which can rightly be called the Namias system as Jerome Namias has led the section since its inception, is one which, compared with the well-tried statistical methods, is scientifically attractive because it refers directly to the circulation of the atmosphere of the northern hemisphere. It is therefore naturally linked with short-range and medium-range forecasting and with the general circulation of the atmosphere. The unit of time is the month – or 30 days – and the first stage in the forecast is to predict the contour-chart at a pressure height of 700 mb, some 10,000 feet above sea-level, as it should appear when averaged throughout the coming 30-day period. Weather patterns averaged over a period of time are of course made very familiar by the climatologists' mean patterns for different months of the year which were illustrated in Chapter 11. The predicted pattern for the coming month may then be regarded primarily as a basis for a comparison with the climatological average and so for a weather-forecast, specifically of temperature and rainfall, by comparison with climatic averages. By extensive statistical methods, particularly thorough for the United States, the most probable anomalies of mean temperature and of total rainfall for the same 30-day period, corresponding with the 700 mb contour-patterns, have been derived making it possible to construct, as hemisphere maps, the distribution of these quantities all over the hemisphere. The predicted rainfall pattern delineates three categories of rainfall: near, above, or below the climatic average. The temperature pattern recognizes five categories, including 'much above' and 'much below'. Given these predicted patterns there can be nothing simpler than to read off the predicted category of anomaly for any region of interest. The forecasts have the great virtue of being clear, definite, and verifiable (checkable) with none of the verbal ambiguities which seem to cling to long-range forecasts. The drawbacks are that the information is limited in scope – referring only to the average temperature and total rainfall for the whole 30 days – and is easily misrepresented by the ill-informed. Enterprising European journalists, when in 1963 they discovered this American source of long-range forecasts, went into print with grossly misleading phrases such as, 'Americans predict 30 days of heavy rain', or, 'Cold weather will continue another 30 days'. They should, one imagines, have known, and perhaps they did, that a month with rainfall over average may yet have many fine days, and that a month which is cold on the whole is not necessarily cold all the time. Pictures speak louder than words but newspapers must have words and experience shows that if a verbal forecast is required the forecaster must choose the words himself if he is not to be misinterpreted. Even then he must expect to be misleadingly paraphrased.

In the above brief account of the Weather Bureau's technique, nothing has been said on the most vital point of all: how is the average chart for the coming month predicted? Actually it is not an easy question to answer because the method seems to reduce in the end to a subjective judgement by the responsible forecaster (who in the past has most often been Mr Namias himself), based on his experience supplemented by quite a number of different and not necessarily consistent indications. The changes in the 30-day mean charts from month to month may to some extent be extrapolated; situations which often persist, and those which do not persist may be recognizable in some degree; the medium-range forecasts for five days ahead are given weight. But the proof of this pudding of many ingredients is in the eating, and it is claimed that the predictions are more right than wrong. Over a test period, the predictions as they applied to the British Isles were found to be little better than chance but, as would be expected, the success is higher for the United States.

The system developed in the Meteorological Office of the United Kingdom under my own general direction is, in some respects, very different from that of the Weather Bureau, but the two have this in common – that they are based on the study of the conditions over a large part of the hemisphere. The issue to the public of long-range forecasts by the Meteorological Office began in Britain only very recently, in December 1963, after a long period – some ten years – of research and trial; and, after careful consideration of the worthwhileness of issuing forecasts which, experience had shown, could be expected to be correct or broadly on the right lines no more than twice out of three times. As, however, the forecasts were in any event being prepared for research purposes and research was to continue the decision to make them available was generally welcome. At the same time the opportunity was taken to put together and place on sale a new bulletin, *Monthly Weather Survey and Prospect*, which gives interesting climatological data and a summary of the weather of the past month as well as the prospect for the coming month. In the middle of the month there is a reassessment of the prospects for a 30-day period extending half-way through the following calendar month, but the climatological data and maps are not prepared for publication at that time.

The methods used in the Meteorological Office in order to arrive at the long-range forecast are various and still exploratory, but the concept which has proved most useful to date is that of the analogue. Put very simply, a search is made through all previous years' records (some eighty years are available) to find occasions when the weather conditions of the immediate past, mainly of the past month, were, for the same time of year, most nearly reproduced. The basic assumption is that similar situations are likely to develop in similar ways and so provide a valid basis for a prediction. There is of course every justification for the assumption provided that sufficiently close analogues can be found covering the essential physical quantities, although the concept has aroused some criticism mainly on the grounds that it is 'elementary' and 'non-scientific'. When, however, one begins to apply the elementary and non-scientific principle to the actual problem the complications begin to gather, the procedures become elaborate, and much of the professional knowledge of the team of forecasters is brought into play. Electronic computing programmes are used to assist in the sorting, and a number of different criteria are used. The first comparison is made with monthly mean patterns of the thickness of the 1000–500 mb layer of the atmosphere over a large part of the hemisphere. This quantity is effectively a measure of temperature and may be translated into terms of temperature. A similar comparison is made with mean surface pressure patterns, and a third with the sequences of synoptic weather situations, especially as they have affected the eastern Atlantic and western Europe. Additional weight is placed on conditions in the past few days and those predicted by short-range forecasting methods for the next few days, the important point being that a past year which failed completely to reproduce this phase could hardly be trusted to give good guidance for a further period in the future.

The final forecast is drafted after a lengthy technical discussion has been held amongst a number of workers who have a special research interest in the problems. Attempts are made to diagnose those characteristics of the current season which seem to be most significant in relation with the general circulation of the atmosphere. Such characteristics may be: abnormal longitudes of the main troughs and ridges in the upper flow patterns, a recurrent tendency to meridional flow patterns, to blocking actions, or to cyclogenesis in normal or abnormal localities. Distributions of sea-surface temperatures and of the extent of the polar ice and the continental snow-fields are also examined, and the attempt is made to relate departures of these distributions from their climatic averages with the current circumpolar circulations.

The forecast is therefore subjective, but experience shows that a concensus of opinion is not difficult to reach when the significance of each piece of evidence has been discussed. The outcome is a verbal forecast for the United Kingdom which gives the mean temperature and total rainfall for the coming month as a comparison with the climatic average, and also offers any other information on trends and special phenomena which seem likely to characterize the month, say a December or a June, regarded as a member of the family of all Decembers or all Junes. It is too early yet to say whether the degree of success is proving to be better or worse than was expected when the series began, but backed up as it is with considerable research effort not only in attempting to refine present techniques but also attacking basic circulation problems by the most fundamental methods available, it is reasonable to hope that the long-range forecasts will claim a permanent position in the public service.

There is nothing like seriously trying to perform a task, if one aims to assess the nature of the problems that arise and the kind of research which is likely to lead to greater success. The combination of forecasting with research in these new ventures, is a valuable feature in both the United States and British organizations.

Chapter 13

The Control of Weather and Climate

THE ability to control his environment – physical and biological – is I suppose the basis of man's successful competition with other forms of life, and the thought of controlling the weather and climate of this planet has an immediate appeal; especially perhaps to those scientists who have been concerned with the great and spectacular enterprises of our generation, the release of tremendous amounts of energy by the conquest of the atomic nucleus, or the expenditure of vast amounts of wealth in the conquest of space. Conquest is an intoxicating pursuit, and the thought that there should, on our own little planet, be any happening beyond our control is quite unacceptable to these imperious and impious conquerors; indeed we can sometimes detect a note of impatience and exasperation in the urgent demand by non-meteorologists for an attack of this untamed climate of ours, the source of endless discomfort and continual disaster. Certainly it does seem absurd that when man has made himself master of the major forms of life on earth, animal and vegetable, and is making good progress with the control of the most elusive micro-organisms, when he can harness the rivers and at least put a wall around the oceans, he can do nothing whatever to control the weather and must be content with protecting himself, and his animals and plants, against its rigours with clothing, housing, heating, cooling, drying, and humidifying and must insure as best he can against the excesses of heat and cold, flood or drought, or destructive storms. However, after the clarion call for a serious attack to subdue this unyielding element must come a sober assessment of the nature of the task with the preparation of a practicable and worth while plan of campaign if this is possible.

My own view, for what it is worth, is that the artificial alteration of weather and climate on the large scale is one of the least promising of objectives, not because it will prove entirely impossible but because what does become possible will not be sufficiently advantageous to be worth while. One might keep it in mind that climate and weather are the manifestations of energy changes on a scale which is gigantic compared with any energies that man can yet dispose of. Energy is likely to become cheaper and in time nuclear sources of energy combined with automation may conceivably bring almost any engineering enterprise within reach, but, not looking too far beyond the horizon, it does seem likely that costs will remain decisive and that it will be more sensible to use artificial energy to keep ourselves warm or cool than to warm or cool the whole atmosphere, to pipe water where we need it rather than to control the rainfall, to travel to the sunshine rather than to bring the sunshine to where we live – indeed, for Mahomet to go to the mountain.

A few numbers may be of interest here:

Population of the earth	$\simeq 2.5 \times 10^9$ persons
Area of the earth's surface	$\simeq 5 \times 10^8$ square kilometres
Mass of earth's atmosphere (2 million tons of air for each person)	$\simeq 5 \times 10^{15}$ tons
Kinetic energy of the winds	$\simeq 10^{14}$ kilowatt hours
Energy of evaporation or condensation of rain (100 million kW hour/year per person)	$\simeq 3 \times 10^{17}$ kilowatt hours per year
Cost of energy of rain at 1 penny/kW hour	$\simeq £10^{15}$ sterling per year
	$\simeq £400,000$ per person
Energy of a large (megaton) nuclear bomb ⎫	$\simeq 10^9$ kilowatt hours
Energy converted in 1 thunderstorm ⎭	
Cost of a thunderstorm at 1 penny/unit	$\simeq £4$ million
Number of thunderstorms on earth	$\simeq 10$ million/year

1 kW hour = 860 kcal = 3,600,000 joules.

We may proceed by listing the factors which seem to have outstanding importance for the climate of Earth in the belief that if we are to alter the climate in any substantial way, we must work through one or more of these factors.

The first is the sun, but I do not think the arrogant space conquerors have any thoughts as yet on altering the activity of this source of all our wealth and comfort. I have no doubt at all that if there were theories on how to make the sun hotter or colder there would be scientific adventurers unable to resist the temptation to make experiments, and it is perhaps comforting to most of us that playing with this particular fire has, so far, no attractions. Taking then the astronomical factors as immutable, the next question is whether we may alter the amount or distribution of energy which is used by the atmosphere in the climatic machine. Recalling that 35 per cent of the heat of the sun that is poured on the earth is reflected to space unused in any way, the prospect arises of altering this 35 per cent, the so-called albedo, one way or the other. The analysis of the reflection shows that only 10 per cent of energy falling on the ocean surface is reflected: the ocean surface appears almost black in space photographs of Earth. This could be whitened perhaps by covering with some kind of foam, the logistics of which may be left over for a moment. The land surface has a highly variable reflective power, from fresh snow almost 100 per cent to dark soil almost zero, so that there are evident possibilities. An obvious major change would be to blacken the snow of the polar caps and the winter continents and so greatly to enhance their absorbing power in the sunny part of the year: if this could be done, a major effect on world climate could be confidently predicted. On the same count, what might be the effect of covering the earth's desert lands with black polythene sheeting I do not know, but it could be striking: the cost of covering, say, the Sahara might be about a billion dollars, but it would not last very long.

The most significant factor in the albedo of the earth is, however, the very high reflectivity of the clouds which cover about half the earth's surface at any time. There seems little prospect of altering the optical properties of these ephemeral features on the large scale, but it is worth examining the possibility of reducing or increasing the average amount of cloud in the sky by interfering with their microphysical properties discussed in an earlier chapter. Although clouds, on average, cover about half the sky, precipitation falls over only say 5 per cent of the earth so that 90 per cent of all the clouds are relatively thin or transient, and the amount might well be altered if clouds could be made to form or clear more rapidly or more slowly, when in this condition. The many different dynamical processes that are responsible for cloud formation and cloud clearance need to be examined to decide whether they are likely to be affected by artificial interference with the micro-physics, presumably by forms of seeding.

A further possibility of interfering with the world heat balance, is to direct the attack to the very high atmosphere where the quantity of air is relatively small and practicable operations may seem less insignificant. All the heat from the sun must come through this region and all the heat lost must pass outwards the same way. In these respects the outer atmosphere may be thought of as a window surrounding everthing within and interference with the window could be important. The first idea may be to pollute it with reflecting dust, having in mind that natural volcanic eruptions in the past have noticeably affected the solar radiation reaching the earth. The most famous was the Krakatoa eruption of 1886 which was said to have reduced the power of the direct sunshine by as much as 10 per cent over a limited range of latitudes and which affected the colours of the sunset and sunrise quite strikingly for many months. Ten months after the volcanic eruption (in February 1963) of Mount Agung on the island of Bali, the colours of the sunset in England appeared exceptionally impressive with an unusually rosy hue, and the effects were still apparent at the time of writing this account, early in 1964. If one were deliberately to use large amounts of nuclear energy to convey dust to the high atmosphere, say between 30 and 60 kilometres aloft, an effect could be observed, but what the change of weather would be in various parts of the world – if there were any detectable change at all – is a question to which the answer is quite unknown.

One might of course be a little more subtle than merely to dirty the window of the atmosphere. It has been suggested that the natural ozone in the atmosphere between 20 and 60 kilometres, which is responsible for the high temperature near the upper limit of this range, would be removed by suitable chemical reactions and so upset the temperature distribution. Effective quantities could be in the order of a ton per square kilometre which approaches a billion tons (10^9) for the whole earth, sufficiently forbidding, but even if it were done the thermal effect, certainly unpredictable, might well be insignificant in terms of weather. Other gases such as nitric oxide and carbon monoxide with suitable absorptive and emissive properties could, if introduced into the high atmosphere, modify the local radiation balance profoundly for a short time, hours or days, but again there is no convincing argument to show that weather would be affected one way or another.

Interference with evaporation from the oceans would be another way of attacking world climate at its foundations, for half the energy put into the atmospherc is used first in evaporating water and is liberated only when and where precipitation falls. On the small scale of artificial reservoirs, molecular thicknesses floating on the surface have been successful in partially inhibiting evaporation and the large-scale problem presents a challenge.

From the fundamental point of view, all forms of interference with the surface boundary condition of the atmosphere need examination. The distribution of land and sea is not likely to be radically altered, but there may be local changes within our resources and having far-reaching effects. A number have been suggested such as the blocking of narrow straits, for example Gibraltar, through which ocean waters and energy are known to pass in very large quantities. Even blocking the Bering Straits between Alaska and Siberia has been suggested as a way of isolating the Arctic Ocean; there are other possibilities. The flooding by sea water of large areas now below sea-level has interesting possibilities, but most of the suitable regions are in desert country and there is reason to suppose that a large salt lake would do no more than perhaps bring rain to a very narrow littoral. Natural lakes in desert areas, for example the Dead Sea, bring little advantage.

The destruction of mountains would permit the freer flow of the winds, and no meteorologist is in doubt that the earth's climate is much affected by the orography so that, once more, it is the size of the task which decides whether action can rationally be undertaken. Finally, on the large scale, one may think of two possibilities; the one to interfere with the major ocean currents, and the other to remove by artificial means the large permanent ice-fields of Antarctica and Greenland – gigantic engineering tasks. It is estimated that the energy of a million megaton nuclear bombs would suffice for Greenland, and the feeling that one is well inside the realm of fantasy is very strong.

The suggestions touched upon so far refer to the artificial modification of the basic climatic controls, in effect the modification of the 'boundary conditions' to the atmospheric circulation problem, and there is another class of possible methods which refer to the internal working of the weather systems. Can we interfere with the mechanisms of the dynamical systems which determine the weather and climate? The most important systems are the showers and thunderstorms, the result of vertical convection and instability, discussed in Chapter 6; the baroclinic depressions with their fronts and rain areas together with the complementary anticyclones discussed in Chapters 9 and 10; the tropical cyclones (hurricanes or typhoons), important mainly for the enormous damage which they are liable to produce; and, finally, the components of the general circulation itself, the trade winds, the monsoon circulations, and so forth. A comment on each of these is called for.

Beginning first with the largest scale, the general circulation, the possibilities of direct interference except through the basic controls already mentioned seem particularly remote. The machine is a colossal heat engine driving against its own internal friction and the power so used exceeds all the sources available to man by many orders of magnitude. If we are to interfere significantly it will not be by pitting our puny strength against the natural forces but by cunningly controlling some one of the working parts.

The baroclinic or frontal depressions are a promising point of entry because theory shows them to be self-developing instability systems which start from small beginnings. Why should we not attempt to control these small beginnings, that is to initiate new depressions where and when we want them without waiting for nature to start things in her own way? It is certainly a very fair question, although preliminary study of the problem does not give much encouragement. In principle, for theoretical discussion, one may think of each new depression as beginning from some minor disturbance, but in fact the atmosphere is so organized that quite large disturbing factors are always present; there is really no lack of triggers to start the mechanism. New depressions do not appear at random, but form in preferred regions related to the topography and to the large-scale flow-pattern at the time, and it is probable that any direct interference, for example by introducing an artificial source of heat, would quickly be swamped by the natural process of development, unless the heat source were comparable with that naturally available in the early stages of development of the depression. The size of depressions, some hundreds of miles in diameter, means that the mass of air involved is of the order 10^{13} tons, millions of millions, and the energy is correspondingly tremendous. To be significant in the company of the natural temperature differences, one might speculate that an artificial source would need to increase the temperature of 10^{12} tons of air by $1°C$, the energy needed then being about 10^{18} joules, the equivalent of 100 megaton bombs. It is a vast amount to throw into the system to try to set off a depression in a rather different position and at an earlier time than it would otherwise have started, knowing that thereafter it would be out of one's control and once more subject to natural forces. There could hardly be a less promising way of disposing of one's wealth.

The tropical cyclone may be rather more promising material. Despite the fact that the way in which they develop is not well understood, their wayward behaviour and devious tracks after they are formed suggests something which may be responsive to cunningly disposed interference. As is well known, many situations which seem to threaten hurricane development fail completely to mature and if, as is likely, some threshold of instability needs to be crossed, it would be tempting to see what a large nuclear bomb could do. But experimenting with nuclear bombs at low levels is not a healthy pastime, and this kind of experimental research would never get the approval of governments unless there were some promise of military value. We may perhaps be grateful that for the moment there is none. There is, however, another possibility. According to certain theoretical studies, a tropical hurricane can be maintained only by drawing continuously upon the energy of the warm tropical ocean waters. If this theory is correct, it would follow that a hurricane might be nipped in the bud and prevented from developing, or at least diverted from some particularly vulnerable place in its track, by cutting off this source of energy in some way. Unfortunately, in the neighbourhood of a hurricane the seas are rough and experimental interference correspondingly difficult, but there may be a possible opening here.

On the smaller scale, the triggering of showers and thunderstorms by artificial heat sources is not wholly out of the question. The heavy rain and hail which occurred with the recent volcanic eruption off Iceland, shows that the process would work if the heat supply were sufficient. It is, however, very easy to see that in most practical situations the power to initiate convection using a large source of heat, even if it were to lead to a shower somewhere in the neighbourhood within half-an-hour or so, would be an utterly useless capability, and it is not at all easy to imagine circumstances in which it would be otherwise, but the possibilities may be borne in mind.

This examination of the scope for interference with the dynamical mechanisms of weather does not encourage optimism – if that is the right word – but there is one other point of entry which seemed at one time to be really promising, that is through the microphysics, the drop-making processes which must go on in clouds if rain is to fall. This fascinating branch of weather science was discussed at some length in Chapter 5, and it may be recalled that two distinct processes are recognized as likely to lead to the formation of raindrops which, if they are to be large enough to fall to the ground as rain, must be about a million times heavier than an ordinary small cloud droplet. The one process requires that a relatively small number of ice particles should form at temperatures well below the freezing-point, the other that a relatively small number of water droplets much larger than the usual size should, for some reason, occur. Both these processes, we believe, take place naturally; but clouds, sometimes thick and extensive, do occur without the production of rain or with less rain than might have been expected, and the question arises whether either process could usefully be stimulated artificially.

Rain-making by magical processes was an essential skill of the wise men, the witch doctors of primitive tribes, and the possibility of doing something can never have been far from the minds of peoples afflicted by scarcities; but it was not until 1946 that serious scientific interest was widely aroused by some remarkable experiments carried out in the United States by Langmuir and Schaefer, and by Vonnegut. It was shown in the laboratory that a volume of cloudy air at a temperature well below freezing-point, that is a supercooled water cloud, could be made to precipitate by the introduction of particles of solid carbon dioxide ('dry ice'), or by the introduction of fine particles of silver iodide. The dry ice, at a temperature of $-70°C$, leaves a trail of ice crystals which quickly capture all the excess water and fall out; the silver iodide particles act as freezing nuclei if the cloud temperature is more than some $5°C$ below freezing-point, and the outcome is the same. Schaefer was quick to translate the laboratory result to the free atmosphere and in one of the most famous experiments in the history of meteorology demonstrated how a natural supercooled cloud, a common enough occurrence, could be stimulated to precipitate by seeding it with solid carbon dioxide released from an aircraft above. The water cloud was transformed in a few minutes to a thin cloud of ice crystals which fell out leaving a clear lane. The amount of precipitation was quite insignificant for practical purposes, but it seemed that the first step had been taken.

From this moment, it may be said that artificial rain-making and artificial cloud-clearing became established as the legitimate objectives of scientific study. The capability of modifying super-cooled clouds, both layer clouds and convective clouds, was confirmed in several countries. At the same time, not surprisingly, rain-making on a useful economic scale was immediately envisaged and in the United States, almost the only country in the world where meteorological services to the community are provided by private enterprise on a significant scale, commercial operators were soon offering to relieve drought: rain-making was a saleable commodity.

The ethics of commerce were never those of science, and no one is particularly surprised if a hair-restorer, nerve tonic, or washing powder fails to live up to the manufacturer's claims. Perhaps, therefore, much of the scathing criticism which has been levelled against commercial rain-makers by disinterested meteorologists is unjustifiable. The methods, so far as anyone could know or demonstrate, might have worked and then the story would have been different. But the fact is that rain-making was marketed mainly in the United States on a big scale without waiting to prove that the processes were successful; even now, after more than ten years of commercial activity, large sums of money are being spent annually by farming communities although very careful assessment of the available evidence by an authoritative body in the United States, supported by the expert opinions of cloud physicists in other countries, leads to the conclusion that no result of economic significance has yet been demonstrated. This unsatisfactory state of affairs lasting for so many years is rather remarkable. It might be thought that the validity of the methods would have been established one way or the other in a few years, but the problem has proved exceptionally intractable. Natural rainfall is so sporadic in its occurrence and so highly variable in its distribution, that it is quite impossible, by any method yet put forward, to distinguish natural rain from the artificial product. If weather-forecasting were a perfect science, artificial rainfall would be readily detected, but there is nothing more difficult to predict than the amount of rain. Whether or not some rain will fall is relatively easy to decide, but rain-makers do not offer to produce rain from cloudless skies. Probably nowhere in the world does a forecasting service claim to be able to forecast the amount of rainfall even a few hours ahead within better than 50 per cent – leaving any amount of latitude for the claims of rain-makers. Tests have, in this unfortunate state of affairs, necessarily rested on statistical evidence. Has the rainfall in a seeded area over a long period been greater than the climatological expectation, taking as control regions neighbouring areas which have been left to nature? In cases of randomized seeding, do the seeded cases show evidence of the greater rain? Where the evidence is available and has been independently analysed, it seems that nothing certain can be deduced. In some cases success which is 'statistically significant' is claimed, but in other cases seeding experiments have apparently caused a 'significant' reduction in rainfall. It is a very confused picture which leads most independent scientists to recommend the suspension of all large-scale experiments and the intensification of basic research, a recommendation which would carry more weight were it not known that independent scientists always call for more basic research – and are probably right. But the view that is gathering more and more support is that natural processes of rain production are after all rather efficient and that almost all the rain which is made possible by ascending motion in the large dynamical systems of depressions and convective cloud does in fact fall out as rain. The non-raining clouds which might respond to artificial stimulation are rarely capable of producing anything of economic significance.

Speculation and comment could continue indefinitely but enough has been said to show the state of the subject. It may be added that local fog has been cleared by a sheer frontal attack, that is by burning it off. FIDO was the name given to keeping an airfield runway clear using petroleum burners on a lavish scale, and there could be less costly, more suitable methods – chemical or electrical. But operations over such a small region are hardly to be described as weather modification. After all, in a sense, the building of a house or the placing of a cloche in the garden is to modify the climate by producing artificially sheltered conditions, and somewhere a line must be drawn. It seems reasonable to regard climatic control as interference with natural weather processes affecting large territories of tens of miles at least, and to look upon smaller scale activity as protective measures against undesirable features; it is in this sense that the term is generally used.

A list of the projects for climatic modification which could receive further examination – scientific, technical, and financial – is then:

1. Alteration of earth's albedo:

   (a) *Oceans:* cover with reflecting white foam.
   (b) *Land:* darken the desert areas.
   (c) *Snow:* blacken the snow surfaces of polar regions and winter continents.
   (d) *Clouds:* reduce or increase amounts of cloud by interference with microphysics, nuclei.
   (e) *High atmosphere:* pollute with reflecting dust.

2. Alteration of radiation balance by introducing various substances into the high atmosphere.

3. Alteration of evaporation from oceans, using suitable substances for covering the surface.

4. Alteration in distribution of land and sea:

   (a) *Reclaiming flooded land* ⎫ effects probably
   (b) *Flooding low-lying land* ⎬ quite local.
   (c) *Blocking of narrow straits.*

5. Reduction of mountains.

6. Diversion of ocean currents.

7. Interference with major dynamical processes of weather.

8. Interference with microphysics of rainfall.

Having made this admittedly rather superficial survey of the possibilities of climate control, it is proper to try to take up some rational position towards future work in the field, but it is not easy. The most notable fact of all is that while numerous schemes, all extremely costly, may be envisaged as certain to have some effect on the climate of the world, our knowledge of general circulation and climatic processes is still too uncertain to permit us to predict with any confidence the outcome in terms of world climate of any one of the schemes. Here I deliberately and categorically discount predictions made, with whatever show of confidence, only by individual workers. Personal predictions, not carrying general scientific conviction, are an excellent basis for harmless small-scale experiments to test them if such experiments can be devised, but are no basis for deliberate large-scale action aiming hopefully at some desired result while putting other interests at risk.

The question can then be put in a rather crucial way: should large-scale experiments be undertaken in order to provide information, or should such actions be deferred until, by research, our knowledge has become sufficiently certain virtually to eliminate the risks at present inherent in experiments? If the latter and more cautious attitude be adopted we may, with some confidence, postpone practical climatic control for a generation and engage in basic meteorological research programmes with such resources as we can command.

If, however, we really wish to make progress quickly, we cannot deny ourselves one of the two greatest weapons in the scientific armoury, namely experiment, and it would be justifiable for representative specialists to get together to study the design of experiments and to weigh the risks. The risks of any large interference with climate are particularly great because vast numbers of the world's peoples live on the subsistence limit, and already suffer tremendous hardships when natural climate runs for a time in an unfavourable trend. Should a large-scale experiment, intended perhaps to improve the climate in some region, have the unfortunate result of producing a worse climate over some highly populated and backward area, the penalty in human suffering could be appalling and the excuse that some well-intentioned group of scientists had made a miscalculation would hardly be adequate. It is not surprising that some meteorologists find the whole area of large-scale weather control quite repellent in our present state of ignorance.

For my own part I should be very well content to take the weather and climate of the world as it comes and leave mankind to make the best he can of his environment, interfering only where the ill-effect on others is demonstrably negligible. We already have cold wars; the prospect of having hot or cold, wet or dry, climatic wars with the innocent neutrals perhaps the greatest sufferer, has no attraction. It is, however, wise to remember that scientists are persuasive people and governments notoriously equivocal. If therefore, we are to avoid the danger of countries independently engaging in large-scale experiments, with credulous governments led by fallible scientists, it would be most advisable to set up an international body to survey all proposals. A world agreement at this stage, to do nothing liable to interfere with weather or climate outside national frontiers without agreement between all countries concerned, would not necessarily be a futile gesture, for experience shows that small groups of specialists even in powerful countries may rush into foolish enterprises which a simple machinery for prior international consultation could have prevented.

While the precautionary creation of a controlling machinery would be a wise present step, I believe we are as yet too ignorant to design any large-scale experiment which would not be too much of a gamble for innocent third parties. But work could be begun, and the requirements of experiments could be examined to lay down some elementary principles. The first is that the objective of any experiment should be stated and published in advance, and the method of determining the outcome of the experiment should be sound and approved by competent specialists. In the limited field of artificial stimulation of rainfall, large resources have been wasted because the necessary observations to determine clearly the result of the enterprises were not assured beforehand. On the world scale, the need is more serious for great issues are at stake. Without extending a somewhat technical argument to the stage of tedium, one may briefly say that if an experiment is designed to produce some change of weather or climate, we should insist that we have some reliable way of knowing what the weather or climate would have been if the experiment had not taken place. Weather-forecasting should be sufficiently reliable and objective, and sufficient observations should be available to establish the facts over all regions of importance. If climate over long periods is in question, careful statistical analysis is necessary beforehand to specify what period and what data are needed to give results carrying conviction. Large-scale experiments in climatic or weather control are likely to be extremely costly, dwarfing the present costs of meteorological observing, data processing, and basic research. It is an elementary request that, by improving the provision of weather data all over the world and extending the effort on research into the general circulation and forecasting, meteorologists should be permitted to prepare themselves for the conduct of wise and economical experiments, and the avoidance of disastrous enterprises.

Arising naturally from considerations of the control of weather and climate is the question whether anything that man has done hitherto has had any effect unintentionally; bearing in mind in asking this question that people generally seem to have the conviction that good weather, whatever that may be precisely, is the natural order of things, and that specially bad weather must have some special cause. What more likely than that some innovation, especially some action that we dislike on other grounds, is the cause of the unwanted weather? By emotional logic of this kind bad weather has been blamed upon anything remotely affecting the atmosphere; railways, gunfire, motor-cars, aeroplanes, and radio have all been suspect characters in their time, and at present we have likely evil-doers in powerful rockets and nuclear explosions. One thing at least can be said against these newer agencies, that they do pollute and disturb the atmosphere on a scale which is at least noticeable locally and for a short time. Nuclear explosions we also know beyond argument, have introduced radioactivity into the atmosphere on a world scale and to a degree which cannot simply be brushed aside as of no practical consequence. Is it not at least reasonable to suppose that the explosions and rockets may in some subtle way have disturbed the weather, perhaps by introducing new substances into the high atmosphere, perhaps nuclei of condensation into the lower atmosphere?

The possibilities have been seriously examined by a number of different national bodies and international organizations such as the World Meteorological Organization and COSPAR, the scientific body for co-operation in space research, and the findings are, so far, all consistent. There is no evidence that weather has been significantly affected except for a short time in the immediate vicinity of an explosion, a matter of a few hours and miles, and there is no good theoretical reason to lead one to expect that the effect would be otherwise. The agent arrested for behaving suspiciously has come through every examination with his meteorological character almost unblemished; he has done no harm to the weather and does not seem to have the capability, whatever other crime he may have committed.

Nevertheless, it is good that public apprehension should be voiced and that responsible authorities should not be allowed to become complacent. There is no reason to doubt the conclusions of experts at the present time, but there is also every reason to believe that the capability of disturbing the climate, although in an unpredictable way, is on its way if not already with us. Large rockets now used for boosting space vehicles may each carry a thousand tons of exhaust gases – mainly water vapour and carbon dioxide – into the high atmosphere, and scientists are rightly afraid that the process could interfere with some of the delicate experiments which it would be interesting to carry out on the outer fringes of the atmosphere. But weather is mainly an affair of the multi-billions of tons of air in the troposphere, and one would need to be very ingenious to introduce anything likely to affect the amount of radiation entering or leaving the system to a detectable degree – short of using nuclear energy deliberately to that end. It is estimated that meteoric dust enters our atmosphere from outer space at a rate of perhaps 1000 tons per day. What effect, if any, this has in climatic processes we do not know but it is against this kind of natural background that artificial interference must be viewed.

Finally, it is remarked once more that to distinguish natural weather from artificial weather by merely examining the observations is almost impossible, and when some indignant sufferer looks for a scapegoat to carry the blame for 'unprecedented bad weather', the climatologist is usually at a loss to find any basis for the notion that the weather did not arise perfectly naturally. To him it will appear just a sample, if rather an unusual sample, taken from the variety of events which go to make up the natural climate, and matched in the records by similar happenings in times when the bomb, the rocket or aircraft, or whatever the artefact under suspicion, had not been invented. Even climates vary by natural processes, a matter which calls for a new chapter.

If one is looking for a criminal agency continuously damaging to our natural climate, it is the pollution of the atmosphere from the incomplete burning of fuels – oil, petrol, gas and solid – and the use of the atmosphere as a convenient reservoir in which to dump the waste products, the smoke, acids, and hydrocarbons. There is a true climatic effect in the loss of sunshine and the creation of smog, and there may be other effects less obvious although far and away the most serious factor is the poisoning of the air we breathe.

In January 1966, while this volume was in the press, some very important reports on Weather and Climate Modification were published in the United States, both by the National Science Foundation and the National Academy of Sciences. These gave excellent surveys of achievements and possibilities, and the conclusions on rainmaking by cloud seeding were more favourable than in previous reports. A striking expansion in research and experimental work was accordingly advocated. One formal recommendation was to raise the direct Federal support for weather modification from the 1965 level of $5 million to at least $30 million by 1970. This would be a tremendous programme in the context of meteorological research, dwarfing the efforts elsewhere in the world, except in the USSR. Important discoveries must surely emerge from investigations on this scale, and if there is still no guarantee of the development of techniques of weather control having real economic importance, the prospect must now be viewed very seriously.

Chapter 14

The Ever-changing Climate

THE mention of 'climatic change' brings first to mind those variations through the remote past, *'Climate through the Ages'*, to quote the title of a remarkable book by the late C.E.P. Brooks, which we associate in particular with the Ice Ages. It is, however, less generally known that however we may choose to divide the earth's long history, in tens of years, centuries, millennia, or spells up to millions or milliards of years, the common characteristic is always change. The climate of the sample of years since 1940 has been measurably and, in a strictly practical sense, significantly different from that earlier in the present century; the climate of the twentieth century has on the whole been different from that of the nineteenth or eighteenth; the climate of the last ten thousand years has been strikingly different from that which preceded it, and the climate of the last million years has been very different from that of any previous period, going back at least some hundreds of millions; and so it goes on. The question at once arises of what do we mean by climate, and all we can safely say is that it is a summary statement of the weather conditions experienced over a period of years, sufficient in length to smooth out the effects of single exceptional years, and that the climate varies with the choice of period. There is no such thing as a 'normal' climate.

There is no part of the earth sciences more fascinating than the study of the changes through the thousands of millions of years since the earth first took shape, but the meteorologist is in a position to look backwards through his observational record and instrumental measurements for a comparatively trivial length of time. Good records from the upper air are available only for a few decades and for a limited part of the earth; practically nothing is known from instrumental observations taking us back more than a century, except here and there a few rainfall or temperature measurements in the few countries of advanced modern civilzation; 200 years may be taken as the limit. Everything we know about climates more ancient than that has been gleaned from other kinds of evidence, from historical records, from archaeology, and further back than a few thousand years from all kinds of geological and geophysical evidence, the physical structure of the earth's crust and the biology of the fossil record. The theories and hypotheses of the changes of climate revealed by the diverse kinds of evidence are correspondingly varied, and open up a range of astronomical and geophysical arguments which takes us far away from the problems of the atmosphere. Even if this is properly a part of weather science, it would be a hopeless task to attempt to present a balanced account in a brief chapter. Brooks quoted Kipling most aptly:

'There are nine and sixty ways of constructing tribal lays,
And – every – single – one – of – them – is – right,'

and went on to say that there are at least nine and sixty ways of constructing a theory of climatic change. It is a labyrinth of science which one enters at one's peril, at the risk of never escaping with one's life, but no account of weather processes can be complete without some kind of historical perspective and even a thumb-nail sketch can be helpful.

The earth itself came into being as a planet to the sun about 5000 million years ago. Whether it began as a hot body detached from the sun or as an accumulation of cold cosmic matter is a question still being debated, but in either case it is thought to have settled down relatively quickly on this scale of time into a planet much as it is today, with a hot interior maintained by the energy of its radioactivity and an outer shell of solid crust, liquid ocean, and gaseous atmosphere, effectively in thermal equilibrium with the sun. During any short period of time, and in this context many thousands of years is a very short period, the changes in the crust, in the land–sea distribution and the orography, were of no significance, but over longer periods – from a million years upwards – they were controlling factors, and it is in the geological record that most of the evidence of past climate is to be found. Even this record has little to say that is unequivocal and relevant to our subject beyond the Cambrian period, some 600 million years ago, and here we may begin.

Perhaps the most surprising fact revealed by this record is that for much the greater part of the time the climate of the earth has been relatively warm everywhere with no permanent ice and relatively small differences between poles and equator.

During three periods in the last milliard years there has apparently been an Ice Age with large areas under permanent ice. The first of these, the pre-Cambrian, affected both hemispheres; the second, the Permo-Carboniferous some 300 million years ago was experienced mainly if not exclusively in the southern hemisphere, while the last which set in only about a million years ago is still with us. When we speak of Ice Ages we do not always have it in mind that we are in one at the present time, for the amount of water locked up in the ice caps of Greenland and Antarctica is prodigious, something like 4 per cent of all the water of the earth, sufficient if melted to raise the sea-level by 300 feet or 100 metres.

There is then a major question: what could have happened to bring permanent ice-fields back to a planet which had been completely clear for countless centuries or to clear the ice once it had accumulated? I do not propose to try to list the sixty-nine varieties hinted at by Brooks, nor even a selection of them, but as a meteorologist merely to note that there is at least one process which would almost certainly suffice. If extensive mountains were created in high latitudes they would become ice-capped and could then act as the firm foundation for a new permanent ice-field. And if the mountains were to disappear, especially if the land were below sea-level, the foundations would collapse and the permanent ice would, it is reasonable to suppose, crumble away and melt in the warmer ocean. It has been claimed, by a Russian scientist, that if the present polar ice-fields were destroyed they would not return under our present conditions – and perhaps the notion of blasting the ice away into the oceans by nuclear explosions was in mind. But I doubt the argument. There is probably still sufficient high ground buried under the ice of Greenland and Antarctica to provide a secure anchorage for permanent ice-fields to form and prosper.

There seems little reason to doubt that this factor, orogenesis, explains much, but whether it is the main reason for the Ice Ages we cannot of course say with confidence. Certainly the specialists are agreed that important periods of mountain building have preceded the main glaciations, and some at least tell us that the denudation of the mountains by the action of glaciers and weathering took place sufficiently rapidly to reduce the relief to comparative uniformity in a sufficiently short period, some millions of years.

On this hypothesis, the essential factor for an Ice Age is high ground in high latitudes which could arise either by new orogenesis or by the movement of existing land into polar regions, consequential to the 'drifting of the continents' or to 'polar wandering'. In the last few years much new evidence in favour of continental drift, that is the movement of the great land masses over the earth's surface, has been provided from paleomagnetism, and new theories of the movements of the earth's crust, associated with convection within the hot interior, have been put forward. Once we admit, as it seems we are bound to do, the possibility of the continents moving from one side of the earth to the other in the course of 100 million years and of the pole finding itself in the middle of the Pacific Ocean (or something related with the present Pacific Ocean) the possibilities for long-term climatic change become legion.

There seems, however, to be some solid ground, so to speak, for the belief that during the last few million years the geography has remained much as it is today on the broad scale, that is apart from some mountain-building and erosion and modifications in the coastlines and shallow seas. The present glaciation followed the period of the Alpine orogenesis of the Quaternary which was perhaps its cause. It would, however, be wrong to get the impression that an Ice Age is a simple event, a period of formation and increase of the ice-fields, of maintenance, and of decrease to final disappearance; whatever the course of events, it is not simple. During the present Ice Age, of the Pleistocene, there have been several waves of advance and retreat of the glaciers and the continental ice-fields. It was only some 10,000 years ago that the permanent ice last departed from the British Isles, a period which seems sufficiently near to modern times to bring a chill wind to the imagination and to tempt even a serious scientist to wonder if another cold wave could come upon us suddenly.

During this post-glacial period of 10,000 years there have been important variations, difficult to summarize because of the complexity but, if we may follow a recent survey by H.H. Lamb, showing four main features. Something like 5000–3000 BC we come to the so-called climatic optimum when average conditions were probably warmer than at any time since. There was much open water in the Arctic, the average European summer was 2–3°C warmer than is normal nowadays and there were summer rains in the Sahara and those parts of the Near East which are now desert. Then, with the peak about 900–450 BC came a cold epoch when continental glaciers advanced considerably and those of the Rocky Mountains south of 50°N first appeared.

The next swing towards a warmer climate attained a secondary climatic optimum in the Middle Ages, say AD 1000–1200. Once more the Mediterranean had a wetter period and at the same time a warmer one, but nowhere was the change more important than in the sub-Arctic regions from Scandinavia to Iceland, Greenland, and the neighbouring coast of North America. This was the period of the early settlements of Greenland and of the remarkable Viking voyages to Labrador and even to Newfoundland, where the grape-vine was then growing. Greenland was colonized in the tenth century in a climate which was suitable for the grazing of sheep and cattle, at a time when coastal ice presented no difficulty to shipping, and the settlements were maintained for some hundreds of years. As time went on, however, the pendulum swung once more and the colonies finally perished, around AD 1400, defeated by the worsening climate and the indigenous ice-tolerant Eskimo. Land which had been cultivated became permanently frozen in depth and remains so to this day; an agrarian community has no answer to 'permafrost'.

The following cold period culminating from say AD 1550–1700 has received the title of the Little Ice Age and was a period of serious stress in lands much farther south than Greenland. It was marked by an exceptionally high frequency of severe winters in western Europe, but since that time, and especially since the middle of the nineteenth century, the trend was towards higher winter temperatures once more, at least until about 1940. As in other periods the changes were most striking, and most important, near the edge of the Arctic where the presence or absence of ice may make a difference of many degrees in summer temperatures. The amelioration of the climate of Iceland and the coasts of the Arctic, bringing to mind the great Viking days, was a matter which came very much into prominence in the years just after World War II, when the free exchange of information revealed many consequences of practical importance: the migrations of the cod fisheries into more northern waters, and the long periods of the year when Arctic ports were remaining open. It is, however, now generally recognized that the optimum had by that time already been passed, at least for a while. The year 1940 is usually taken as the time when the trend reversed and colder winters became more frequent once more. The very exceptional winter of 1962–3, the coldest in Britain for 150 years, seemed to complete the evidence so to speak and cause serious-minded and responsible people to ask whether we were swinging into a severer phase of the Ice Age once more. The last three or four years have been peculiar in another way: the rainfall of tropical Africa has suddenly jumped, many lakes have risen, including Lake Victoria, and new precautions may need to be taken. At the same time the Great Lakes of Canada have become short of water and Britain has had the driest spell of three years in a century. Has something new arrived? Will it last? At present we can only speculate.

There is a curious feature about this climatic record which is worth a remark. Looking backwards through time we have remarked upon turning points about AD 1940, AD 1800, AD 1100, 450 BC, 3000 BC, and 10000 BC and may observe that the interval grows with the time elapsed. We may be very sure that this is merely an illusion due to the fact that as we approach the present time the evidence grows in detail and that short-period fluctuations, such as are lost in the distant perspective of thousands of years ago, are obvious and even dominant landmarks in the recent record. In other words, we are mixing up the time scales of events when we compare the Ice Age with the Little Ice Age, or the climatic optimum with the short-lived maximum of the early part of the twentieth century. Variations on many time scales are going on continuously and simultaneously and probably with as many different causes.

It is this mixing of the scales of time which makes the discussion of climatic change even more confusing than it need be, in terms both of the empirical evidence and of the likely theoretical explanations. One thing we may be virtually certain about is that the alternations between the Ice Ages on the scale of many millions of years or the fluctuations within the present Ice Age on the scale of tens or hundreds of thousands of years, are quite irrelevant to practical politics or to economic planning when a hundred years is a very long time. The significant climatic amelioration of the last 200 years and the set-back, which may or may not be continuing, of the last twenty years are, on the other hand, events of real moment which should be exercising the minds of practical people. But when on the basis of a few severe winters there is talk of a return of the Ice Age, or on the basis of a hot dry summer there is talk of a rapid return to a 'Mediterranean climate' for north-west Europe, it is only an example of that scientific sensationalism which is always associated with ignorance and, unfortunately, is not always discouraged by scientists who should know better.

The meteorologist must, however, keep one common factor in mind. All changes of climate, on whatever scale, must have meant changes in the general circulation of the atmosphere, and in the positions and intensities of the main climatic zones. The way to understanding is, therefore, by a better understanding of the general circulation, how it is controlled, and how it may vary. This is the basic problem of weather and climate to which we inevitably return, whatever the original interest may be.

Chapter 15

Weather Science of the Future

IF the writing of this book has been at all successful, it will have given the impression of a modern developing science, both intellectually exciting and economically rewarding, ranging over an impressive diversity of ideas and techniques, and having numerous growing points and loose ends. There are many books on scientific subjects written for the general reader which seem to have a different objective, which give attention almost exclusively to the discoveries and positive achievements of the science, which are tightly packed with specialist information and are avidly read by the many enviable people with long memories and an insatiable interest in knowledge for its own sake. For my own part, I can think of no less sordid reason for writing a book than to provide an account as reliable as one's state of knowledge permits for the guidance of those who delight in being up to date, but this need not be the only reason for a book on science nor are those acquisitive of technical knowledge the only readers one may have in mind. There are others who are actively repelled by specialist knowledge presented *ex cathedra* for the enlightenment of the uninitiated and have no sympathy with science, a body of tedious if possibly useful knowledge, or with scientists who write with such dogmatic authority. It seems then just as important to give the more valid impression of the scientist as one who works always on the frontier between knowledge and ignorance, and of science as the exploration of the unknown. There can be nothing more unknown than the future progress of science, dependent as it must be on discoveries, by their nature unpredictable, so that to speculate on the subject will form a suitable ending to an inconclusive book.

The feature of meteorology which is most characteristic, and most costly in both money and manpower, is the world network of regular daily observations, and this is a good point of entry for a view of the future. Historically, the observing stations have been established piecemeal to meet pressing national or international aviation requirements, with the World Meteorological Organization providing the machinery for essential co-ordination. Quite recently, however, it may indeed be dated from the WMO Congress of 1963, the concept of a World Weather Watch as an organic entity has been accepted. As yet it is an idea, or an ideal; no international fund is available commensurate with the size of the task and there is no definite promise of adequate support; but WMO in 1963 set its special fund at the level of 1·5 million dollars, created a special section within its secretariat and appointed an Advisory Committee of twelve scientists* to assist its deliberations on this and other scientific developments. All this is earnest of good intentions and one is therefore justified in assuming that sooner or later what should be done will be done. One may therefore predict that the world will, in the course of a score of years or perhaps a little longer, be covered with a sufficiently close network of observations to give a fair approximation to a complete picture of the weather, including the clouds, winds, and temperatures, everywhere and at all times.

One cannot, in the nature of the case, foresee entirely new techniques, but present thinking would call for radio-sonde balloon stations at a density of 1 : 50,000 square miles over all land areas and all oceans where sufficiently populated islands are available. The wide expanses of oceans will, in part, be filled in by fixed ocean weather-ships, although these are particularly costly to maintain, in part by making balloon observations from selected ships of the merchant services along the shipping lanes of the world, in part by using drop-sondes from high-flying aircraft, and in part by balloons drifting with the wind for thousands of miles at predetermined heights. With reliable data from the upper air obtained in these ways to serve as a firm framework for quantitative weather analysis, the picture will be filled in first by observations of clouds from artificial earth satellites, permitting all active storm centres to be located, and, additionally, by surface manned and

* The author of this book is a member of the Committee.

unmanned automatic stations, by radar networks and other auxiliary ways.

Communications must of course expand in parallel, but we are only at the beginning of a phase of revolutionary advances in the speed and capacity of communications systems, and one feels confident that the collecting and distributing of weather-data will present no problem more difficult to overcome than that of actually making the observations. Furthermore, automatic high-speed data-processing will allow any selection, digest, or representation of data to be made for local purposes anywhere on earth.

The provision of immediate weather information to the interested customer, with short-range forecasts as required, will be arranged for all parts of the world and must be in ever-growing demand. The needs of agriculture, industry, commerce, and transport may be expected to evolve along foreseeable traditional lines, and new needs will become pressing as the expanding wealth of the world leads to ever-expanding travel. It is of course in the backward but advancing countries that weather services will soon be most needed for commercial efficiency – soon, rather than immediately, for a country must reach a certain stage of organization before it is possible to take advantage of weather knowledge. Some international scheme for providing for the needs of the large numbers of sovereign territories, many in tropical latitudes, must come if the logic of the case carries any weight whatever. International or world-supported advisory weather centres may be expected with confidence – they are so clearly needed.

The advanced countries also have a long way to go before the application of weather knowledge to practical affairs reaches saturation, for in very few cities of the world is it yet possible for anyone interested in some special aspect of the weather to obtain the kind of helpful advice which it would be technically possible to provide at a cost which would be trivial in any serious context. Experience in the United Kingdom with local Weather Advisory Centres in the large cities has been a revelation, and in due course such institutions must become as normal as travel agencies. Probably in most countries, progress on these lines is retarded by the government monopoly of weather services, combined with the difficulty of collecting payment for minor services in large numbers. It has always been difficult for a democratic country to expand public utilities to the degree called for by the public interest, when the cost in taxation is always evident and tangible while the benefits fall unevenly and remain uncosted; but ways are usually found when there is no vested interest in active opposition, as in the present case. Probably no large city will in twenty years' time be without a local civilian weather-centre open to the public.

The increase in leisure in the affluent age ahead must add to the interest in weather as people become even more mobile, travelling large distances by land, sea, and air, often pursuing some outdoor interest. Knowledge of weather will enhance the pleasures of recreation as certainly as it does the efficiency of commerce, and public weather services are in no way threatened by an age of plenty.

Automation in this profession, as in others, must reduce the need for repetitive work while objective mathematical calculations aided by electronic computing will displace the professional 'weather-forecaster', but the weather-consultant who can match the product to the user will be in growing demand, and for many many years to come the expert scientist will find research problems calling for his attention.

What progress may we look for in the science itself, in the solution of outstanding problems? Weather-forecasting for short periods ahead, a day or two, will be improved by the development of more accurate methods of solving the physical problem by calculation, and of course, the forecasts will be available for all parts of the world. But there is at present no basis for supposing that the improvement in accuracy will be revolutionary within a generation: the basic difficulties deriving from instabilities in weather processes appear formidable. Extended forecasting in general terms for a month or a season ahead will in twenty years' time be a normal service, and the accuracy will be sufficient to make them useful. The forecasts will probably be made by objective quantitative methods and be prepared by high-speed computers as an extension in generalized terms of the short-range forecasts, but there remain good scientific arguments which would justify one in taking a conservative view of the likely gain in accuracy of long-range predictions; the problem is not just mysterious, it is scientifically complicated and intractable.

In twenty years' time there will be few among the old problems of weather for which an acceptable scientific solution is not available in the sense that theoretical models will have been formulated in terms of the basic laws of physics. There are many outstanding problems at the present time which, if this prediction is to come true, must yield to attack. A quantitative dynamical model is still required for most meteorological phenomena: the cumulus cloud and the thunderstorm, cloud systems and patterns of many kinds, some revealed for the first time by artificial satellites; the frontal complex – warm front, cold front, or occlusion – still dynamically obscure although familiar concepts for the past forty-five years; the large-scale bands and areas of rain and cloud, some associated clearly with fronts, others not so: these problems do not seem too difficult for successful attack.

On the larger scale, the evolutions of the depressions and anticyclones of extra-tropical latitudes are now largely explicable from basic theory, and only refinements in understanding remain to be made; but the theoretical analysis of the general circulation of the atmosphere as a whole, and the elucidation of the weather variations from month to month and year to year, has barely begun. It is here that perhaps the most spectacular progress must be looked for in the course of the next generation.

Featuring prominently in modern meteorological research programmes is the study of the high atmosphere above the weather, and it is a facile prediction to say that twenty years will see remarkable progress in describing and understanding the new phenomena of these regions in the stratosphere and beyond, but it is less easy to say how much, if anything, these studies will have taught us about the weather near the ground where we live. At this point we encounter one of the most intriguing and significant questions of meteorology: how much do events in the high atmosphere affect those near the surface of the earth? So far as we can say at present, little of value in more mundane weather science may come from high atmosphere studies, but to know the nature of the interactions – important or otherwise – will be highly significant for the progress of understanding and this we may expect.

The interaction of the atmosphere with its lower boundary, the oceans, the ice-fields, and the solid earth, including the exchanges of momentum, energy, and substance – especially water

– is a field of study which has been worked over in a general way for many years so that the physical principles and processes are in a sense well understood, but it is a far cry from the formulation of principles to the possibility of accurate calculations of events in real time. The transfer problems are as important to the oceanographer as to the meteorologist, and an integration of effort may confidently be expected. It is a curious thing that it is, as yet, impossible to make direct measurements in ordinary natural conditions of any of these exchanges, so that the onus falls upon the theoretician to devise reliable methods of calculation from quantities such as wind, temperature, humidity, and radiation which can be measured; much preliminary work has been done, but there is much that is outstanding to keep researchers busy for many years. Until the nature of the fluxes at the upper and lower boundaries of the weather-bearing atmosphere are understood, the problem of the circulation of this atmosphere cannot be fully formulated; the needs are now well recognized and progress must follow.

There are three distinct ways in which weather science can be of practical value. The first is by the application of knowledge of meteorological quantities and processes to other activities, as we may apply our knowledge of turbulence to the design of aircraft, of winds and waves to the design of ships, of radiation to the heating of buildings, of humidity conditions to the storage of foodstuffs or equipment, of averages and extremes of rainfall and evaporation to the planning of water conservation – the list is endless. This may be called applied atmospheric physics and applied climatology, and will in the future come to employ the greatest number of meteorologists. The second main value of weather science comes from forecasting, from the ability to look ahead in a characteristic and unique way and we have already alluded to its expected development in this direction. The third is weather modification or climatic control, to which a chapter has been devoted but which must be brought forward again to take its place in this sketch of likely progress in the coming years. At every hand when meteorologists advocate the study of their science, as it is their proper place and duty to do, the possible reward of being able to control the weather is held out as a glorious prospect which no wise planner could overlook. To be sceptical about progress in his science calls pity on a man for lack of faith or vision, if not obloquy for disloyalty, so that few voices are heard to discount extravagant speculations. It is even said, and said openly, that a scientist is entitled to overstate his case to attract funds and brains to his pursuit, although it is hard to see why deliberately to mislead is less reprehensible in a scientist than in a salesman. In any event, there are such weighty reasons for studying and advancing weather science for the benefit of humanity, that there is no need to go beyond what sober judgement suggests and to promise the control of weather and climate which it seems likely would bring more trouble than benefit to the world. As I read the present signs, I should say that in the next twenty years no progress of any practical significance will have been made towards controlled interference with natural weather processes, excluding here the creation of local artificial climates by extensions of protective buildings or by direct heating or air conditioning, which have been with us ever since man took advantage of shelter and learned to control fire. I have no doubt that something could be done now to upset the climate of the earth in an unpredictable way if the energy available in nuclear explosions were diverted to polluting the atmosphere or disfiguring the landscape, but there is every reason to believe that no country will embark on mad adventures with no more certain outcome than is likely to be foreseeable for many years to come. There is security and strength in the Advisory Committee set up by the World Meteorological Organization and the Committee on Atmospheric Sciences set up by the International Union of Geodesy and Geophysics acting for the International Council of Scientific Unions. On these bodies, scientists as knowledgeable as can be found will speak with a voice which experience suggests is likely to be heard everywhere, and at present there is unanimity that the time is not ripe for large-scale experiments aimed at altering the climate. Sober judgement suggests that it may never be ripe, but never is not a sensible word to use in this context and if one wishes to imagine wise men in some future time, much wiser scientifically and politically than we are today, controlling the climate of planet Earth to the advantage of its inhabitants it is not entirely fantastic.

Finally in this chapter, I would like to allow myself a comment on the future organization of scientific work in meteorology and especially on the place of the universities and other research laboratories, for in many countries, including the United Kingdom, all is not well and could be better at no greater cost. Weather science has grown up rapidly since World War I with the advance of aviation and at an accelerated pace since World War II, but almost everywhere as a state monopoly – with the United States as a notable exception. In some countries, particularly Germany, meteorology was established early as a university discipline and scientists qualified in the subject could be recruited; but in the United Kingdom and elsewhere, expansion of services ran ahead of education and the gap has not been closed to this day. The universities did not turn out trained meteorologists; the services were therefore compelled to recruit those qualified only in the basic sciences, particularly in mathematics and physics, and to provide the professional and specialized instruction within the establishment. With such a system in being, working successfully, the demand upon the universities for meteorologists failed to appear, and the normal relationship between university and profession never emerged in meteorology. Were the profession much smaller there could be no grounds for uneasiness for university departments would probably not exist; were the profession much larger and less of a monopoly, the normal pattern would probably evolve automatically – we fall between two stools.

Something should be done, and perhaps will be done, in years not far ahead, by making special arrangements to meet these special circumstances where university meteorology is too precious to abandon and yet too limited to be viable in the normal way. A solution which has worked in some countries is to bring university and profession into close partnership so that they work not in the normal relationship of a training establishment and post-graduate research institute to a profession and government research laboratory, but in physical combination with joint research programmes and co-ordinated and complementary teaching and training programmes. Here is not the place to work out a scheme in detail, but there is a peculiar genius in the atmosphere of university studies and university research, which a science and profession learns to live without at its peril. At the same time, the training of recruits to a profession provides an objective and a stimulus without which a university department can become an empty shell. In the developing countries, many too small to support effective weather services and certainly for a long time incapable of creating research laboratories, there is an overwhelming case for international research institutes, perhaps supported by the United Nations; and if these arise, as they seem likely to do, association with university departments will be just as valid for them. The future of education, training, and research in weather science is a matter of the greatest interest – in some ways it is the most interesting problem of them all.

Suggestions for further reading

THE following is a careful selection of some recent and up-to-date books written by scientists who are well-known experts in their subjects. The first list is of books intended primarily for the general reader with scientific interests. The second consists of more technical or professional works intended for serious study but nevertheless in large part within the scope of the non-specialist.

For the general reader

Climate and the British scene Gordon Manley. London (Collins), 1952. 314 pages. 25s.
A fascinating treatment of British climate and weather; ideal reading for the countryman.

The ways of the weather J. S. Sawyer. London (Adam and Charles Black), 1957. 97 pages. 12s. 6d.
A popular little work, mainly on problems of weather forecasting.

Cloud study: a pictorial guide F. H. Ludlam and R. S. Scorer. London (John Murray), 1957. 80 pages, 73 plates. 12s. 6d.
A quite outstandingly interesting discussion of clouds in scientific terms, illustrated by a fine selection of photographs.

The upper atmosphere H. S. W. Massey and R. F. L. Boyd. London (Hutchinson), 1958. 333 pages. 63s.
Mainly an account of explorations of the high atmosphere with rockets and satellites, written by leading workers in 'space research'.

Everyday meteorology A. Austin Miller and M. Parry. London (Hutchinson), 1958. 270 pages. 30s.
Eminently readable, descriptive meteorology.

Exploring the atmosphere G. M. B. Dobson. Oxford (Clarendon Press), 1963. 188 pages. 21s.
Mainly about the upper atmosphere. Excellent general reading.

Where no birds fly Philip Wills. London (Newnes), 1962. 141 pages. 21s.
An account of gliding experiences by a famous exponent of the art.
Contains much about the atmosphere not found in textbooks.

The challenge of the atmosphere O.G. Sutton. London (Hutchinson),
1962. 227 pages. 21s.
A simple but authoritative account of some modern problems by the
Director-General of the Meteorological Office.

The English climate H.H. Lamb. London (English Universities Press),
2nd Edition 1964. 212 pages. 12s. 6d. paperback. 21s. cloth bound.
Excellently written – will interest everyone.

A colour guide to clouds Richard Scorer and the late Harry Wexler.
Oxford (Pergamon Press), 1964. 63 pages, 48 colour plates. 12s. 6d.
A remarkable collection of pictures, expertly described, at a bargain
price.

Rather more technical but mostly suitable for the non-specialist

The restless atmosphere F. K. Hare. (Hutchinson's University Library),
London, 1953. 192 pages. 8s. 6d.
A systematic but not too technical treatment of dynamic climatology.

Hurricanes I.R. Tannehill. London (Geoffrey Cumberlege), 9th Edition, 1956. 308 pages. 36s.
A classical work, mainly descriptive, about the destructive hurricanes
of the West Indies and North America.

Handbook of Meteorological Instruments London (HMSO)
Part I. Instruments for surface observations. Reprinted 1961. 458
pages. 45s.
Part II. Instruments for upper air observations. 1961. 209 pages.
25s.
An authoritative technical description of modern instruments within
the scope of school physics.

Physics in meteorology A.C. Best. London (Pitman), 1957. 159 pages.
18s.
A useful indication of a selection of classical meteorological problems
suitable for students of physics.

Farming weather L.P. Smith. Edinburgh (Nelson), 1958. 208 pages. 15s.
The author is a leading agricultural meteorologist of his generation
and his book deals with the problems of farming with authority and
scientific discipline.

Weathercraft L.P. Smith. London (Blandford Press), 1960. 87 pages.
9s. 6d.
An unusual but successful little book suited to science students in
secondary schools.

British weather in maps J.A. Taylor and R.A. Yates. London (Macmillan), 1958. 256 pages. 21s.

A first-year university course for geographers: mainly descriptive.

Glossary of Meteorology Boston (American Meteorological Society), 1959. 638 pages. $12.00.

Meteorological Glossary Compiled by D.H. McIntosh. Meteorological Office, London (HMSO), 1963. 288 pages. 32s. 6d.

Two similar works containing definitions of most technical meteorological terms.

Meteorology for glider pilots C.E. Wallington. London (John Murray), 1961. 284 pages. 25s.

Although written for a special purpose, this is an unusually lively account for anyone interested in watching the weather.

Clouds, rain and rainmaking B.J. Mason. Cambridge (CUP), 1962. 145 pages. 22s. 6d.

An excellent introduction at the standard of university physics with interesting descriptive matter.

Physics of lightning D.J. Malan. London (English Universities Press), 1963. 176 pages. 25s.

Excellent reading, requiring a basic knowledge of physics.

The flight of thunderbolts Sir Basil Schonland. Oxford (Clarendon Press), 2nd Edition, 1964. 182 pages. 30s.

Lightning theory, damage and protection; interesting historical section.

Understanding weather O.G. Sutton. London (Penguin), 1964. 235 pages. 3s. 6d.

A useful introduction for the student.

Index